W9-DFL-162

Women in Engineering and Science

More information about this series at http://www.springer.com/series/15424

Jill S. Tietjen

Engineering Women: Re-visioning Women's Scientific Achievements and Impacts

 Springer

Jill S. Tietjen
Technically Speaking, Inc.
Greenwood Village, CO, USA

ISSN 2509-6427   ISSN 2509-6435   (electronic)
Women in Engineering and Science
ISBN 978-3-319-40798-2   ISBN 978-3-319-40800-2   (eBook)
DOI 10.1007/978-3-319-40800-2

Library of Congress Control Number: 2016945373

Printed on acid-free paper

This Springer imprint is published by Springer Nature
The registered company is Springer International Publishing AG Switzerland

In memory of my parents—Manuel and Bernice M. Stein

Preface

For almost 40 years, I've said "I'm an engineer" when asked about my career. As to what I do, I explain: when I am traveling on an airplane and I look down and see the lights—that is what I do. I help ensure that this country has electricity. It has been a tremendously satisfying career choice for me.

Did I always know that I wanted to be an engineer? Absolutely not. I entered the University of Virginia (UVa) in the fall of 1972 as a Mathematics major. No one, not even my Ph.D. engineer father, had suggested engineering as a career for me. And there weren't very many female role models in my hometown of Hampton, Virginia, or at UVa. UVa had only admitted women as undergraduate students in the fall of 1970 (under court order). I was in the third class of women admitted and the first class that didn't have a cap on the number of women that could be admitted. Shortly after the start of my college career, however, I did discover engineering, and I transferred to the School of Engineering and Applied Science.

After I graduated I found out that there weren't very many women in the engineering field. Of course, there hadn't been many women engineering professors, undergraduate women engineering students, or graduate women engineering students at UVa, but I thought that was a function of the exclusion of women at the University prior to 1970. I didn't know it was a characteristic of the engineering field in general.

Fortunately for me, my first employer, Duke Power Company (today Duke Energy), sent me to do on-campus recruiting. At a card table in a gymnasium at North Carolina State University in Raleigh, North Carolina, at a career fair, I found the Society of Women Engineers. Through that organization, I began to research historical women in engineering and science and nominate technical women for awards. One thing led to another led to another. Twice, I have been at the White House as my nominees received the National Medal of Technology and Innovation from the President. Today, I tell the stories of great women across all fields of endeavor, but my first love is telling the stories of technical women and, particularly, women in engineering.

This book provides an overview of the development of the engineering field and describes women's contributions. In most history books (not just history books about engineering), women's accomplishments are invisible or marginalized. Not here. I firmly believe that a culture that values women and recognizes their accomplishments is a better society for all of its members.

Come join me and discover engineering history and women's engineering history.

Greenwood Village, CO, USA   Jill S. Tietjen, P.E.

Acknowledgements

As with any undertaking of this sort, many people provided support, assistance, and encouragement. A huge thank you to the following and I apologize in advance for any errors or omissions.

To Alexis Swoboda, who planted the original seed. I am forever indebted.

To all of my Society of Women Engineers colleagues (whom I consider a family unit), especially Beth Boaz, Yvonne Brill, Sherita Ceasar, Jane Daniels, Patricia Eng, Jamie Ho, Gina Holland, Helen Huckenpahler, Suzanne Jenniches, Peggy Layne, Dorothy Morris, Islin Munisteri, Anne Perusek, Mary Petryszyn, Carolyn Phillips, Ada Pressman, Nancy Prymak, Mary Rogers, Meredith Ross, Anna Salguero, Sandra Scanlon, Kristy Schloss, Nanette Schulz, Jackie Spear, Alexis Swoboda, Mary Ann Tavery, Robin Vidimos, Kitty Wang, and Jere Zimmerman.

A debt of gratitude to Betty Reynolds, Kendall Bohannon, Wendy DuBow, Carol Carter, and Sande Johnson, for paving the way.

In particular, Tiffany Gasbarrini and Rebecca Hytowitz at Springer US, for their belief, support, and encouragement.

To my village (Enid Ablowitz, Anna Maria Larsen, Glo Martinez, Nancy Mayer, Pam McNish, Colleen Miller, Marie Sager), without whose love, support, guidance, and at times, redirection, are appreciated more than they know.

To David, who supports all of my many endeavors.

Jill S. Tietjen, P.E.

Contents

Contents

Chapter 1
The Early Days of Engineering

Abstract Although the formal profession of engineering originated in the fifteenth century as a military endeavor, scientific concepts applied in the form of engineering projects have been in existence for many thousands of years. By the eighteenth century and with the emergence of the Industrial Revolution, engineering moved away from purely military applications, into the civilian sphere (hence the term "civil engineer") and began to resemble the various types of engineering we now know. This expansion was made possible by the realization that common principles applied not only to the building of catapults, but to the building of roads and bridges as well. At first, engineering resembled the craft traditions where daughters and wives had been welcomed as apprentices and unpaid artisans. However, as the Industrial Revolution spawned mass production and increased demands for technical education, the sphere became almost entirely male. European universities did not admit women. The engineering schools established in the U.S. in the 1800s also did not admit women. In spite of these barriers, women did manage to contribute to engineering and science. Profiles are provided for scientific and engineering women from the early centuries through the 1800s.

A Short History of Engineering

Although the formal profession of engineering originated in the fifteenth century as a military endeavor, scientific concepts applied in the form of engineering projects have been in existence for many thousands of years [1, 2]. The pyramids of ancient Egypt were begun as early as 2630 B.C. [3]. The Great Wall of China was built prior to 200 B.C. [4]. The Incas at Machu Picchu were excellent civil engineers. During 1450–1540, they designed extensive irrigation and other water-handling systems [5]. The Rialto Bridge in Venice, Italy, crosses the Grand Canal and was built in the sixteenth century [6].

Although not many records of historical women scientists and engineers survive to the present, we know that they did exist throughout antiquity. Early women engineers and scientists include Tapputi-Belatekallim, an early chemical engineer and perfume maker in Babylon circa 1200 B.C.; Pythagoras's wife Theano (c. 500 B.C.), who ran the school after his death; botanist Artemisia of Caria; physicist and

© The Editor(s) (if applicable) and the Author(s) 2017
J.S. Tietjen, *Engineering Women: Re-visioning Women's Scientific Achievements and Impacts*, Women in Engineering and Science,
DOI 10.1007/978-3-319-40800-2_1

philosopher Arete of Crete; marine zoologist Pythias of Assos; Miriam the Alchemist (also called Mary or Maria) circa 200 B.C. who invented laboratory equipment including the three-armed still and waterbath (still found in modern laboratories and known in French as *bain-marie* and in Spanish as *baño de Maria*); and Hypatia (A.D. 370–415). One of the best remembered of the early women scientists, Hypatia invented the astrolabe (a device for measuring the positions of celestial bodies), an apparatus for distilling water, a hydrometer (a device for measuring the density of liquids), and a planisphere [1, 7]

When the term "engineering" first came into use, it described the design of mechanical devices for warfare. Universal scientific principles were applied, for example, in launching projectiles and determining approximately where they would land. The use of scientific principles, whose development was significantly enhanced during the Industrial Revolution, proved far superior to the previous methods, which were basically trial and error [1].

By the eighteenth century and with the emergence of the Industrial Revolution, engineering moved away from purely military applications, into the civilian sphere (hence the term "civil engineer") and began to resemble the various types of engineering we now know. This expansion was made possible by the realization that common principles applied not only to the building of catapults, but to the building of roads and bridges as well [1].

Engineering continued to evolve as the physical world became better understood. Sir Isaac Newton (1642–1727) was led to his greatest discovery—the theory of gravity—by the fall of an apple in 1666. His theory is crucial to physics (a building block of engineering) and forms a key basis for mechanical engineering. His many discoveries affected almost every area of the physical world, with special emphasis on experimental and theoretical physics, as well as chemistry and applied mathematics. Newton invented virtually the entire science of mechanics, and most of the science of optics. He also invented the mathematics he needed, including what is now known as calculus—a basic requirement for all engineering students [8, 9].

Using the mathematical foundation laid by Newton, other key principles underlying the various disciplines of engineering were discovered during the 1700s, 1800s, and into the 1900s. The Bernoulli equation, which described the motions of fluids, was developed around 1730 [10]. The law of electromagnetism, which describes magnetic forces exerted by electrical currents, was formulated in 1820. Many of the basic principles of thermodynamics came into being in the 1820s and 1830s, and precipitated the development and enhancement of steam engines. Electricity and electrical engineering were significantly advanced with electrical generation machinery (1884) and the transformer (1891), which allowed our system of electrical power to develop. Radioactivity was discovered in 1896 and the theory of relativity was put forth in 1905, allowing the development of nuclear power and a wide range of medical applications, including X-ray machines [9, 11, 12].

At first, engineering resembled the craft traditions. Craft and merchant guilds in the thirteenth and fourteenth centuries had established rules for training apprentices to take over what were usually family businesses, and traditionally had welcomed daughters as apprentices and wives as unpaid artisans. Often women carried on the business in the case of a husband's death until a senior journeyman could take over. Sometimes she carried on the business herself indefinitely [1]. However, once the fields that had traditionally been "crafts" were upgraded and transformed into "science," women were no longer welcome. In addition, the Industrial Revolution spawned mass production and increased demands for technical education [1]. Mass production needed large numbers of engineers—apprenticeships did not turn out large numbers nor was it suitable for complex technical training. In addition, home-based crafts could not compete with mass production in price or quantity [13]. As mass production took place in public, it became almost completely male. Women were expected to exist in their own sphere, which was invariably personal, private, and domestic [1].

Furthermore, as the engineering discipline evolved, formal schools were established to teach the needed curriculum. European universities, however, did not admit women. Most European universities had been founded to teach theology, medicine, and law; these professions were all closed to women. So women, with very few exceptions, were denied access to a university-level education. The École Nationale des Ponts et Chaussées (National School of Bridges and Roads) was established in 1747 in France as the first formal school of engineering [1]. In the nineteenth century, the practice and focus of engineering was significantly expanded with the development of Newtonian mechanics and the development of the steam engine. These developments led to an even greater need for formal education [14].

Male French engineers assisted the colonies during the American Revolution, stayed to construct the new nation, and provided the foundation of engineering faculty for the U.S. Military Academy (USMA) at West Point [2]. When it was established in 1802 to educate engineers, the USMA did not admit women. The Rensselaer Polytechnic Institute, established in 1824 in Troy, New York, now the oldest surviving non-military engineering school in the U.S. also did not admit women [1, 13]. Prior to the Civil War, there were only six schools of engineering across the entire U.S.; none admitted women [13]. Despite the odds, a few women did manage to attend engineering schools. The first woman to graduate from an engineering degree program in the U.S. was Elizabeth Bragg, who completed a B.S. in civil engineering at the University of California at Berkeley in 1876 [14].

Elizabeth Bragg was a true pioneer woman in the field of engineering. In many professional fields, talking about the "pioneers" means talking about people who lived in the 1800s. In the field of engineering, however, many of the pioneering female engineers lived and worked in the 1900s. And, most of the earliest female engineers were not formally educated as engineers. Pioneering female engineers needed two advantages to achieve lasting recognition: outstanding achievements and strong personal ties with men in their field [15]. Close personal relationships with husbands, fathers, or other family members enabled women engineers to reach a level of professional acceptance that most other women were denied; it also gave them access to knowledge of particular fields [1]. Let's explore the history of engineering education for women in the early years of the U.S.

1700s

A young woman cannot and ought not to plunge with the obstinate and preserving strength of a man into scientific pursuits, so as to forget everything else. Only an entirely unwomanly young woman could try to become so thoroughly learned, in a man's sense of the term; and she would try in vain, for she has not the mental faculties of a man. —Author unknown [15]

There is little historical data concerning women in engineering during the Colonial period in the U.S., possibly because there were so few of them and because the discipline itself was so young. Women who might have aspired to an engineering career were likely to come across a number of stumbling blocks, much the same as women in other male-dominated professions did. Chief among these stumbling blocks was a fundamental lack of access to formal education. However, the stereotypes about gender-appropriate occupations seems to have been even more sharply defined in science and engineering than in other fields [14].

Exclusion of women from education—and particularly any type of higher education—fit with the mores of the Colonial period. The laws and customs that took hold first in the original thirteen colonies were based on English common law. According to English common law, women's social status was acquired either by birth or marriage. Women had many duties befitting their station but few, if any rights; they were trapped in a condition known now as "civil death [16, 17]." Women were deemed subordinate to men and were expected to play a subservient role, first to their fathers and then to their husbands [16]. Women's sphere was narrowly defined as domestic, and the female role specifically defined as wife and mother—roles later characterized as the "cult of true womanhood." Women were expected to be submissive and possess the values of piety, purity, and obedience. At this time in U.S. history, pursuing any form of rigorous education was considered inappropriate for women; they might harm their reproductive capabilities—especially if they filled their heads with radical ideas [18].

While minority women were no doubt involved in some form of "science" and "technology" long before, during, and after the Colonial period, there is scant documentation to confirm their contributions. With the exception of limited reports of African-American women who engaged in medical practices on the plantations, there is little else to suggest that scientific pursuits were within the purview of minority women until long after the Civil War [1].

The 1800s Prior to the Civil War

Education was not a widespread privilege (or requirement) for many years after our nation's founding. During Colonial times, education beyond the elementary level was denied to females. Formal education was rare, public schools as we know them today did not exist, and what education was available was not free. In fact, to be formally educated—especially beyond the high school level—was a rare privilege throughout most of Western history and exclusively granted to affluent males. When the Constitution was adopted, few colonials had attended school; the literacy rate for white women was about 40 % and for white men, about 80 % [17, 18].

By the end of the eighteenth century, ideas about female education reform were beginning to draw attention. Judith Sargent Murray, an early advocate of educating women as well as men, protested the lack of equality in education between boys and girls in her 1790 tract *On the Equality of the Sexes* [16, 17, 19]. Female private schools, female seminaries or "dame schools," had begun to spring up in the 1780s, although attendance was usually limited to the well-to-do. Most of these schools focused on domestic subjects to ensure that their graduates would attract proper suitors [20].

Women needed to lead the charge to ensure that education was available to women. Two of the more prominent of these women were Emma Willard and Mary Lyon. Credited with being the first person to make secondary education available for women, Emma Willard was able to inspire the citizens of Troy, New York to raise enough money to build the Troy Female Seminary in 1821. More than anyone else, Willard wrought the basic revolution in the nation's attitude toward the education of women between 1819 and the 1830s [15, 21]. Education reformer Mary Lyon established Mount Holyoke Seminary (later College) in 1837. She successfully endowed a school for women that not only exists today, but flourishes, and stands as a testament to her efforts [15, 21].

Also in 1837, Oberlin College set a milestone in education by becoming the first institution of higher education to admit women and students of all races. Oberlin had been established in 1833 as a seminary for men, but later became a college. Women were viewed as a "civilizing influence" on the men and, at first, were actually not allowed to take the same course load as their male colleagues due to their "smaller brains." By 1841, however, women were allowed to obtain the same bachelor's degrees with the same coursework as the men [16, 18].

During the 1840s through the 1860s (the so-called "Age of Reform"), women fought for change in many areas. During this period, female activists sought abolition, the right to vote, equal rights, and educational opportunities [16]. Soon, their efforts began to yield results.

By 1850, most cities had public schools—at least one for girls and several for boys. The state of education for minorities did not yet measure up to even these standards. And it took the better part of the nineteenth century to expand the free education system for males from elementary schools though high school. By 1860, there were only about 40 schools that qualified as high schools in the entire country [16, 18].

Colleges had been established as early as the 1600s in the Colonial states, and by the 1800s were fairly common in the eastern U.S. Young men had an opportunity to attend Harvard College and other Eastern all-male institutions. Although some of the early female seminaries called themselves "women's colleges," they did not measure up academically to these Eastern all-male institutions. However, they did lay the foundation for the establishment of Antioch in 1852, Vassar in 1865, Smith and Wellesley in 1875, and Bryn Mawr in 1885 [15, 16, 18, 21]. None of the women's colleges had an undergraduate engineering program [2].

The picture was a little different in the Western U.S., where most colleges were state-supported and usually coeducational from the time of their founding because males were not enrolled in sufficient numbers to support them, and taxpayers would not support them unless their daughters could enroll. Many of these institutions came about as a result of the Morrill Act of 1862 that has been credited with democratizing higher education and providing colleges for the industrial classes. It also led to more schools that offered engineering education and more engineering programs open to women [13]. By 1870, Wisconsin, Michigan, Missouri, Iowa, Kansas, Indiana, Minnesota and California had established coeducational state universities. The number of women going to college increased dramatically between 1860 and 1920, as educational opportunities became available and women saw the economic and personal benefits of becoming educated [18, 22].

Key Historical Women

The contributions of those few engineering and scientific women of whom we have knowledge anywhere in the world prior to the Civil War are fascinating and in many cases, enduring. The following brief biographies provide a flavor of the lives they led, the accomplishments credited to them, and the difficulties they encountered.

Si Ling-Chi or Lei-Tzu or Xilingshi (c. 2640 B.C.E.)

This Empress of China is credited with discovering how to remove silk threads from silkworm cocoons, thus spawning the silk manufacturing industry in China. Si Ling-Chi directed the development of the silk cultivation and weaving industries [23].

Miriam the Alchemist (First or Second Century A.D.)

Born in Alexandria, Egypt, Miriam was also known as Mary, Maria, and Miriam the Prophetess or Miriam the Jewess. Her major inventions and improvements included the three-armed still or tribikos, the kerotakis, and the water bath. The original purpose of the inventions was to accelerate the process of metals transmuting into gold, but now they are used extensively in modern science and contemporary households. The tribikos was an apparatus for distillation, a process of heating and cooling that imitated processes in nature. Sponges formed a part of the mechanism and served as coolers. The kerotakis was an apparatus named for the triangular palette used by artists to keep their mixtures of wax and pigment hot. The water bath, also known as Marie's bath (*bain-marie*), is similar to the present-day double boiler [24].

Cleopatra the Alchemist (c. Third Century)

Cleopatra was an alchemist who probably lived in Alexandria, Egypt. Her two surviving papers include drawings that show a two-armed still and a type of chemical apparatus. These are believed to be the earliest surviving drawings of chemical apparatus. Cleopatra was interested in weights and measurements. Today, the work of the alchemists is recognized as the forerunner of modern chemistry [25].

Hypatia (Circa 360–415)

The daughter of Theon, a well-known mathematician in Alexandria, Egypt, Hypatia was raised by her father to be a "perfect human being"—in spite of the fact that she was a daughter and not a son. Raised to seek knowledge, she was educated in the arts, sciences, literature, philosophy, and all manner of sports. After her mathematical knowledge surpassed that of her father, she was sent to Athens to study. When she returned to Alexandria, she became a teacher of mathematics and philosophy. Hypatia wrote a number of treatises in algebra including significant information on cones being divided by planes. Her inventions included a plane astrolabe, a device used for measuring the positions of the stars, planets and the sun, and to calculate time and the ascendant sign of the zodiac; an apparatus for distilling water, a process used for distilling sea water that is still used today; a graduated brass hydrometer for determining the specific gravity (density) of a liquid; and a hydroscope, a device used to observe objects that lie far below the surface of the water. Her brutal murder led to the end of the formal study of mathematics in Alexandria for over 1000 years [7, 26].

Queen Sonduk (or Sondok) (Seventh Century)

The Queen of the Silla Kingdom (Korea) for 15 years (approximately 632–647), Queen Sonduk ruled while the oldest observatory in East Asia was built. Located in Kyongju, South Korea, the Ch'omsongdae observatory represents the advances made during the Silla era that included new inventions and discoveries in astronomy, meteorology, engineering, printing, and ceramics [25].

Emilie de Breteuil du Châtelet (1706–1749)

The Marquise du Châtelet was tutored as a young woman because her parents thought her homeliness would prelude her being suitably married and wished to make her single life more tolerable with a good education. However, Emilie, grew into a beautiful young woman of intelligence and wit who not only married, but had a well-known series of lovers including a very long-standing relationship with Voltaire. Her significant intellectual interests were in physics and mathematics.

In collaboration with Voltaire, she wrote *Eléments de la philosohie de Newton* (1738), which explained Newtonian physics for a French audience. *Institutions de physique*, published in 1740, originated as a physics textbook for her son and included principles from Newton and German mathematician Gottfried Leibniz. By now, students were arriving to study with Châtelet and she began the culmination of her life's work, a two-volume translation of Newton's *Principia* into French. It was published in 1759, 10 years after her death, which occurred a few days after the birth of her fourth child [7, 26].

Laura Bassi (1711–1778)

Laura Bassi, an Italian physicist, was fortunate to live in Bologna, Italy, a city that prided itself on being a leading center for women in education. At the age of 20, she was presented with membership in Bologna's Academy of Sciences, which was part of the Institute of Sciences. Shortly thereafter, she received a doctorate in philosophy from the University of Bologna. Although she was offered a chair in philosophy at the University and named a university professor, because she was a woman, she was allowed to give public lectures only by invitation. She was able to overcome significant resistance from various circles in Italian society as her career advanced via the support of her husband as well as members of the academic community, church leaders, and political figures.

She published scientific papers as a result of some of her research. Topics included: air pressure (1745), solutions for problems in hydraulics (1757), the use of mathematics to solve trajectory problems (1757), and bubbles formed from liquids in gas containers (1791). Much of the rest of her research, which did not result in publications, involved fluid mechanics, Newtonian physics, and electricity. Because of her social position in Bologna society, she was also required to write poetry for community events and contribute to literary publications [24].

Maria Gaetana Agnesi (1718–1799)

In 1748, Italian Maria Agnesi published a two-volume, 1020-page manual titled *Analytical Institutions* that significantly enhanced the mathematical and scientific knowledge of the day. The volumes, intended as a textbook for her younger brothers, included analysis of finite quantities (algebra and geometry) in volume one, and differential and integral calculus (analysis of variable quantities and their rates of change) in volume two. Her clarification of the work of the best known mathematicians and scientists of the day, including Leibniz, Newton, Kepler, Galileo, and L'Hopital, was recognized for its importance around the world and translations were sought by scientists and mathematicians. All of this from a woman born in Milan, Italy, into a society where most young women, even in the upper classes, were not even taught to read.

The French Academy of Sciences described her work on infinitesimal analysis as "organized, clear, and precise" and authorized translation of her second volume from Italian into French in 1749. The English translation was published in 1801. In 1750, she was named honorary chair of mathematics and natural philosophy at the University of Bologna, although she never lectured. After her father's death in 1752, she gradually withdrew from mathematical and scientific activities, apparently because she associated those activities with him and the strong encouragement and support he had provided to her in her endeavors [7, 26].

Sophie Germain (1776–1831)

A young woman so determined to study mathematics that she persevered even after her parents made sure her bedroom was without light or fire and was left without clothes so that she would have to stay in bed, Sophie Germain was not allowed into the École Polytechnique at 18 years of age to continue her studies because she was a woman. Her parents had finally relented and allowed her to study mathematics during the day, and she was ready for more advanced education. Undeterred by the refusal of the École Polytechnique to admit her, she studied on her own through notes obtained from other students. She wrote to French mathematician Joseph Louis LaGrange under a pseudonym, and he was so impressed with her comments that he met with her and commended her observations. She later also communicated with German mathematician Carl Friedrich Gauss who was so impressed that, in 1831, he was successful in having the University of Göttingen award her an honorary degree.

In 1816, it happened that vibrations and their patterns was the subject of a competition for the French Academy of Sciences as the mathematical theory to explain them had never successfully been developed. Germain's work, the only entry in the competition, was awarded the grand prize. The mathematics for the vibration patterns are used in the construction of tall buildings, such as skyscrapers, today. Now that she had a prize, she was allowed to attend sessions of the Institut de France. In the 1820s, Germain became interested in number theory and developed a theorem in support of Fermat's Last Theorem, her most important work in number theory. Her theorem has since been generalized and improved, but not replaced [26].

Mary Fairfax Somerville (1780–1872)

One of the first honorary women members of the Royal Astronomical Society (Great Britain), Mary Somerville was responsible for "popularizing" science— writing books and papers that explained science to general readers, many of them women. After spending a year at a boarding school when she was 10 years old, Somerville developed a thirst for reading and arithmetic. She taught herself Latin, and then algebra, after seeing strange symbols in a ladies' fashion magazine. When her parents found out about her interest in mathematics, her father forbade her to study, worried that mental activity would harm her female body. After her first husband died and left her with a modest inheritance, she openly educated herself in trigonometry and astronomy. Most of her friends and family did not support her educational efforts.

Somerville married her first cousin in whom she found someone to support her pursuits of educational and intellectual matters. In fact, her husband encouraged her to expand her studies beyond mathematics and astronomy to Greek, botany, and mineralogy. In 1834, she published *On the Connexion of the Physical Sciences* which presented a comprehensive picture of the latest research in the physical sciences. Her 1831 book, *Mechanism of the Heavens*, contributed to the modernization of English mathematics. Sommerville was occasionally criticized for her "unwomanly" pursuit of science. Nevertheless, she was referred to, both in England and abroad, as "the premier scientific lady of the ages" [24].

Ada Byron Lovelace (1815–1852)

The daughter of the English poet Lord George Byron, Ada Lovelace now has a computer language named (Ada) after her. A somewhat sickly child, Lovelace was tutored at home and was competent in mathematics, astronomy, Latin, and music by the age of 14. Totally enthralled by Charles Babbage's Difference Engine (an early computer concept), at 17 years old, she began studying differential equations. As proposed, his second machine, the analytical engine, could add, subtract, multiply, and divide directly and it would be programmed using punched cards, the same logical structure used by the first large-scale electronic digital computers in the twentieth century.

In 1842, the Italian engineer, L.F. Menabrea published a theoretical and practical description of Babbage's analytical engine. Lovelace translated this document, adding "notes" in the translation. Her notes constitute about three times the length of the original document and, as explained by Babbage, the two documents together show "That the whole of the development and operations of analysis are now capable of being executed by machinery." These notes include a recognition that the engine could be told what analysis to perform and how to perform it—the basis of computer software. Her notes were published in 1843 in *Taylor's Scientific Memoirs* under her initials, because although she wanted credit for her work, it was considered undignified for aristocratic women to publish under their own names. Ada Lovelace is considered to be the first person to describe computer programming [7, 26].

References

1. Ambrose S, Dunkle K, Lazarus B, Nair I, Harkus D (1997) Journeys of women in science and engineering: no universal constants. Temple University Press, Philadelphia
2. Barker A (1994) Women in engineering during World War II: a taste of victory, unpublished, Rochester Institute of Technology
3. Rediscover Ancient Egypt with Tehuti Research Foundation. www.Egypt-tehuti.com/phyramids.html. Accessed 26 Sept 2000
4. All About the Great Wall of China. www.enchantedlearning.com/subjects/greatwall/Allabout.html. Accessed 26 Sept 2000
5. Wright K, Kelly J, Zegarra A (1997) Machu Picchu: ancient hydraulic engineering. J Hydraulic Eng 123:838–840
6. Grand Canal (Italy). http://Encarta.msn.com/index/conciseindex/. Accessed 26 Sept 2000
7. Alic M (1986) Hypatia's heritage: a short history of women in science from antiquity through the nineteenth century. Beacon, Boston
8. Sir Isaac Newton and chronology. pp 1–2. www.reformation.org/Newton.html. Accessed 1 July 1999
9. Wolfson R, Pasachoff JM (1987) Physics. Little, Brown and Company, Boston, MA
10. Daniel Bernoulli and the making of the fluid equation. http://pass.maths.org.uk/issue1/bern/index.html. Accessed 1 July 1999
11. A gallery of electromagnetic personalities. www.ee.umd.edu/~taylor/frame8.htm, frame3.htm, frame4.htm, and frame7.htm. Accessed 1 July 1999
12. Tesla N. www.neuronet.pitt.edu/~bogdan/tesla/bio.htm. Accessed 1 July 1999
13. Turner EM. Education of women for engineering in the United States 1885-1952. (Dissertation, New York University, 1954), UMI Dissertation Services, Ann Arbor, MI
14. Kass-Simon G, Farnes P (eds) (1990) Women of science: righting the record. Indiana University Press, Bloomington, IN
15. Rossiter MW (1992) Women scientists in America: struggles and strategies to 1940. The Johns Hopkins University Press, Baltimore, MD
16. Flexner E, Fitzpatrick EF (1996) Century of struggle: the women's rights movement in the United States, Enlargedth edn. The Belknap Press of Harvard University, Cambridge, MA
17. Baer JA (1996) Women in American Law: the struggle toward equality from the new deal to the present, 2nd edn. Homes & Meier, New York, p. 15
18. Harris B (1978) Beyond her sphere: women in the professions in American history. Greenwood Press, Westport, CT
19. Garza H (1996) Barred from the bar: a history of women in the legal profession. Franklin Watts, New York
20. Dexter EA (1950) Career women of America: 1776-1840. Marshall Jones Company, Francetown, NH
21. Webster's Dictionary of American Women (1996) SMITHMARK Publishers, New York
22. The land grant system of education in the United States. http://www.ag.ohio.state.edu/~ohioline/lines/lgrant.html. Accessed 11 Apr 2000
23. Jones C (2000) 1001 Things everyone should know about women's history. Doubleday, New York, NY
24. Shearer BH, Shearer BS (eds) (1997) Notable women in the physical sciences: a biographical dictionary. Greenwood Press, Westport, CT
25. Proffitt P (ed) (1999) Notable women scientists. Gale Group, Farmington Hills, MI
26. Morrow C, Perl T (eds) (1998) Notable women in mathematics: a biographical dictionary. Greenwood Press, Westport, CT

Chapter 2
Women Can Be Engineers, Too!

Abstract Prior to the late 1800s, engineering education was available only to male students. For most women whose aspirations were inclined toward science or engineering, the educational system and associated opportunities would not be available until late in the twentieth century. Thus, many of the early women "engineers" were not educated as engineers in the sense one would expect today. In 1893, the official records only documented three women as having received engineering degrees in the U.S. As women did endeavor to be educated and practice as engineers, a backlash developed. Once educated, women wanted to participate as men did in the engineering societies established for camaraderie, professional development, and networking opportunities. Those societies did not welcome women. It would take many years for the engineering societies as well as the engineering honor society to admit women. Profiles of engineering and scientific women from the late nineteenth and early twentieth century are provided.

Engineering Education in the Nineteenth Century

Prior to the late 1800s, engineering education was available only to male students. While the U.S. population was centered in the East, the colleges in the West and mid-West formally admitted women earlier than East Coast institutions, primarily because many state-supported institutions were established as a result of the 1862 Morrill Act. But even then, the number of women formally studying engineering in the late 1800s were very few. For most women whose aspirations were inclined toward science or engineering, the educational system and associated opportunities would not be widely available until late in the twentieth century.

Some women were able to slip in through the cracks that were starting to show in the male-dominated bastions of engineering educational institutions, either as students enrolled in engineering curriculum or in related science curriculum. Many of the early women "engineers" were not educated as engineers in the sense one would expect today.

Ellen Henrietta Swallow Richards was one of these non-traditional engineers. Although she was not an engineer by training, she contributed much to the establishment of the forerunners of environmental and sanitary engineering and is

© The Editor(s) (if applicable) and the Author(s) 2017
J.S. Tietjen, *Engineering Women: Re-visioning Women's Scientific Achievements and Impacts*, Women in Engineering and Science,
DOI 10.1007/978-3-319-40800-2_2

credited as the woman who founded ecology and home economics. When she applied to the chemistry department at the Massachusetts Institute of Technology for a graduate degree in chemistry in 1870, she was not accepted because the department did not want its first graduate degree to go to a woman. She was not rejected, however, (as she was at all the other universities where she had applied), but instead was allowed to enroll as a candidate for a second bachelor's degree. She was classified as a special student who did not have to pay tuition (she had already received a bachelor's degree from Vassar College). Richards did not know that MIT had admitted her without tuition so that they could deny she was officially enrolled, if anyone complained. She completed her work for a doctoral degree, but MIT refused to grant it to her. MIT formally admitted women in 1878 [1, 2].

Elizabeth Bragg became the first woman to obtain an engineering degree. She graduated in civil engineering from the University of California at Berkeley in 1876 [1]. Kate Gleason was the first woman to enter Cornell's Sibley College of Engineering in 1884, but did not stay to complete her studies, as she was called back to help the family business [4]. Perhaps the second female engineer by education, Elmina Wilson graduated in 1892 from Iowa State College with a civil engineering degree and was the first female instructor at that school [3]. In 1893, Bertha Lamme graduated from The Ohio State University with a degree in mechanical engineering with an option in electricity. She was the first woman to graduate with a degree in a field other than civil or architectural engineering [1].

When the Society for the Promotion of Engineering Education (later named the American Society for Engineering Education) was formed in 1893, only the three women noted above were recorded as having received engineering degrees in the U.S. However, as women began to enter the educational system, graduate, and then try to find work as engineers, a backlash developed [3].

Professionalization

Professionalization of engineering began in the late nineteenth and early twentieth centuries. As women were finally able, at least in small numbers, to gain an engineering education and engineering employment, they also endeavored to join the engineering societies. These organizations, however, did not welcome women and developed a strict hierarchy of requirements for each of several levels of membership. They were already in the midst of upgrading themselves and the entrance of women into the profession was not seen as a positive development by most of the men involved in the leadership positions of these organizations [2].

Professionalization, in this case, meant upgrading the membership or image of a profession by excluding or diminishing the influence of persons who could be perceived to be "amateurs." Professionalization in engineering included deliberately creating barriers between engineers with college degrees and relevant professional experience and those other "engineers" who had learned their jobs by experience and lacked the "professional" credentials. In professional societies, professionalization often meant raising the standards of membership and led to great concerns about the perceived prestige of the organization. As most engineering schools did not admit women (and thus women could not get the desired "professional" credentials), the most significant impact of professionalization was to exclude women [2].

Professionalism was probably also a by-product of the state of engineering education. Engineering educational standards in the late 1800s and early 1900s were not yet at a level necessary to earn a college education. As a consequence, engineers were not invited to serve on national scientific advisory boards, nor were engineers recognized as part of the established scientific community until 1916. And not until 1932 was the Engineering Council for Professional Development (ECPD), now known as ABET, created to provide accreditation of engineering degree programs, in partial response to reports sponsored by the engineering societies [4].

Professional Societies

By the end of the nineteenth century, civil and mechanical engineering were firmly established as engineering disciplines, with electrical and chemical engineering following closely behind [1]. Engineering societies were forming. Organizations, such as these engineering societies, are deemed the hallmark of a profession. These societies define intellectual style and norms of conduct and generally act to promote the interests of their members. The early engineering societies placed a high value on free enterprise, individualism, hard work, ambition, and success—characteristics of a rugged male culture, with concepts of prestige, status, and professionalism closely intertwined with masculinity [4].

The "Founder" societies, the five original engineering societies that founded the United Engineering Society in 1904 (which later became the United Engineering Foundation) included the American Society of Civil Engineers (ASCE), the American Institute of Mining Engineers (AIME) (now called the Society for Mining, Metallurgy and Exploration), the American Society of Mechanical Engineers (ASME), the American Institute of Electrical Engineers (AIEE—a predecessor organization to the Institute of Electrical and Electronics Engineers—IEEE), and the American Institute of Chemical Engineers (AIChE) [5–7]. A brief look at their history and their admittance (or more accurately, their lack of admittance) of women shows the impact that professionalization and the associated membership requirements had on the recognition for and advancement of women in the engineering profession.

ASCE, America's oldest national engineering society, was founded in 1852. Twelve founders met at the Croton Aqueduct in New York City on November 5, 1852, and agreed to incorporate as the American Society of Civil Engineers and Architects (later the ASCE) [8, 9]. Emily Warren Roebling, probably the first woman field engineer, became the first woman to address the ASCE in 1892, when she argued that her husband should not be replaced as the formal director for the construction of the Brooklyn Bridge [1]. In 1909, Nora Blatch de Forest, a graduate of Cornell University in the top five of her class, was admitted as a "junior member" of the ASCE but was unable to advance any higher. When her junior membership expired in 1916, ASCE refused to promote her to full membership, in spite of her meeting the stated requirements, and instead, dropped her from the rolls. She brought a lawsuit against the Society, but did not prevail [2]. Elsie Eaves became an associate member of ASCE in 1927 and later the first female member (1957), first female life member, first female Fellow, and the first woman elected to honorary membership (1979) [1, 10].

AIME was founded in 1871 by 22 mining engineers in Wilkes-Barre, Pennsylvania [7, 11]. The first woman member was Ellen Henrietta Swallow Richards, who became the first woman member of any engineering society when she was elected a full member of AIME in 1879. Richards was aided by her MIT degree, her publications in mineral chemistry, and the fact that her husband was a vice president of the organization [2]. In 1917, the Woman's Auxiliary to the AIME was established and is still active today as the WAAIME [11, 12].

By 1880, 85 engineering colleges had been established in the U.S. and most offered a full mechanical engineering curriculum. Thirty engineers met in New York City in February 1880 and decided to form the ASME. In April, a formal organizational meeting was held with 80 engineers at the Stevens Institute of Technology in Hoboken, New Jersey. The first annual meeting of the organization was held in November 1880 [13]. The first woman member, Kate Gleason, was admitted to full membership of the ASME in 1914 [14]. Lydia Weld, a 1903 naval architecture graduate of MIT, became an associate member of ASME in 1915. She was allowed to become a full member in 1935, when the ASME finally granted full membership status to women [3].

By 1884, twenty-five prominent figures in electrical technology signed a "call" to establish an American electrical national society, mindful that civil, mining, and mechanical engineers had already established their own national societies. Twenty-five electrical engineering practitioners met in the headquarters of the ASCE on April 15 to devise an organizational structure for what became at first the AIEE. The first general meeting was held on May 13, also at ASCE headquarters [15]. In 1926, Edith Clarke, who would later become one of the first AIEE female fellows, was the first woman to address the AIEE [16]. As late as 1942, there were only three women in the AIEE and over 17,000 men [16]. The Fellow grade was established in 1912 for engineers who had demonstrated outstanding proficiency and had achieved distinction in their profession. It was not until 1948, however, that the first women were elected AIEE Fellows [17]. These three distinguished women were Edith Clarke, who significantly contributed to knowledge about and modeling for electric utility systems; Vivien Kellems, the founder of Kellems Company, a manufacturer of cable grips and shell lifters; and Mabel Rockwell, who significantly contributed to electrical control systems [18–20].

AIChE was founded in 1908 at the Engineers' Club in Philadelphia by 19 men. Chemical engineering was just coming into its own and was somewhere between chemistry and mechanical engineering. The founding of AIChE helped to establish chemical engineering as a separate discipline. At the time, the founding members of

AIChE believed that about 500 people were practicing chemical engineering across the country [21]. The first female member of AIChE, Margaret Hutchinson Rousseau, the first woman to receive a Sc.D. in chemical engineering from MIT and who made significant contributions to the field, was not admitted until 1945 [17].

In addition to the Founder Societies, an engineering honor society, Tau Beta Pi, was established. The engineering equivalent of Phi Beta Kappa (which had been founded at the College of William and Mary in 1776), Tau Beta Pi, was established at Lehigh University in 1885. It was founded to:

> ... mark in a fitting manner those who have conferred honor upon their alma mater by distinguished scholarship and exemplary character as undergraduates in engineering, or by their attainments as alumni in the field of engineering, and to foster a spirit of liberal culture in engineering colleges.

Membership in Tau Beta Pi was limited to men until 1969. Women's badges had been authorized in 1936 as an alternative to membership for women. Only 619 Women's Badges had been awarded by 98 chapters until women were admitted to full membership in 1969, 84 years after the founding of the organization [22–24].

Other engineering societies, in addition to the Founder societies, came into existence in the early to mid 1900s. However, these organizations also excluded women from membership or often relegated them to lower membership status. Marie Luhring was elected as an associate member of the American Society of Automotive Engineers in 1920 [14]. That same year, Ethel H. Bailey became the first full female member of the organization [14]. The American Society of Safety Engineers, founded in 1911 in New York City with 62 members as the United Society of Casualty Inspectors, admitted its first female member, Vera Burford, in 1946 [17, 25]. The first time a woman was elected as a junior member of the Society of Naval Architects and Marine Engineers was 1946 [26]. The first female fellow of the Illuminating Engineering Society, Gertrude Rand, a researcher on the way color perception is affected by illumination and on color blindness, was elected in 1954 [16, 17].

Lillian Gilbreth, "the first lady of engineering" and the co-founder of the field of industrial engineering, was made an honorary member of the Society of Industrial Engineers in 1921 (as a personal favor to her husband, Frank Gilbreth), but not admitted to regular membership [2, 27]. She was, however, the first woman elected to the National Academy of Engineering (NAE), an event that occurred in 1965, only one year after the founding of the NAE in 1964 [28, 29].

Early Twentieth Century

The struggle for women to enter the engineering profession made some little progress in the early twentieth century. Women are known to have graduated in engineering from some universities, even if only in ones or twos. Finding a job was the next problem. Attaining the credentials necessary for recognition as even a junior or associate member of one of the professional engineering societies was a further obstacle faced by most women. And then, if a woman married, she was expected, except in very rare cases, to become a wife and mother and abandon all thoughts of a career. The situation was so dire that *American Men of Science*, 1921 edition, lists zero women as employed in engineering [2]. The 1920 Census, however, reports that of 130,000 engineers counted, 41 were women, up from 21 in the 1890 Census [30].

Key Women of This Period

The key women in engineering whose most significant contributions occurred after the Civil War and prior to World War I, were generally not educated in "engineering." With admission to engineering programs prohibited for women in almost every instance, most of the women who impacted the engineering profession either were educated in other scientific fields or gleaned their "engineering" knowledge through on-the-job training.

Ellen Henrietta Swallow Richards (1842–1911)

Ellen Swallow was admitted to MIT as a "special student" and earned a bachelor's degree there in chemistry in 1873 (the first woman graduate of MIT) after having graduated from Vassar (as one of its first graduates) in 1870. She was denied an earned doctoral degree from MIT, as the school did not want a woman to be the first person awarded a doctorate in chemistry. While a graduate student, she executed a complete survey of Massachusetts drinking water and sewage for the Massachusetts Board of Health (1872), taking more than 40,000 samples. Through this work, she warned of early inland water pollution. She also contributed the first Water Purity Tables and the first state water quality standards in the U.S. From 1873 to 1878, she taught in the MIT chemistry department without a title or salary as the first women teacher. She also did extensive research in mineral analysis.

After her marriage in 1875 to Professor Robert H. Richards, head of the department of mining engineering at MIT, she persuaded the Women's Education Association of Boston to contribute the funds needed for the opening of a Woman's Laboratory at MIT. As assistant director to Professor John M. Ordway, an industrial chemist, Richards began her work in the laboratory by encouraging women to enter the sciences and to provide scientific training to women. In 1879, she became the first woman member of the American Institute of Mining Engineers. She was certainly technically qualified for this membership classification; however, her husband's status of vice president of the organization contributed significantly to her selection.

By 1883, the laboratory had proved so successful that MIT allowed women to enroll in regular classes and closed the laboratory. Richards' work in the laboratory had resulted in several books and pamphlets including the seminal *Food Materials and Their Adulterations* (1885). This publication influenced the passage of the first Pure Food and Drug Act in Massachusetts. Her work included analysis of air, water, and food and led to national public health standards and the new disciplines of sanitary engineering and nutrition. The interaction between people and their environment, her areas of study, have led to Richards being called the founder of ecology.

In 1884, she was instrumental in setting up the world's first laboratory for studying sanitary chemistry. She served as assistant to Professor William R. Nichols in the new laboratory and held the post of instructor on the MIT faculty for the rest of her life. From 1887 to 1889 she supervised a highly influential survey of Massachusetts inland waters.

Since 1876, Richards had been on the forefront of promoting education for women, especially in science. In 1881, Richards helped found the Association of Collegiate Alumnae (later renamed the American Association of University Women). In 1882, she helped to organize the science section of the Society to Encourage Studies at Home.

After 1890, she concentrated most of her efforts on founding and promoting the home economics movement (at first it was called domestic science)—an achievement for which she is primarily noted (and frequently criticized for its detrimental effect on women's equality). Home economics was given definition by a series of conferences held in Lake Placid, New York, organized and chaired by Richards starting in 1899. She was involved in the formation of the American Home Economics Association (1908) and was appointed in 1910 to the National Education Association [1, 2, 16, 31–34]. Richards has been inducted into the National Women's Hall of Fame.

Emily Warren Roebling (1844–1903)

Emily Warren Roebling, generally considered the first U.S. female field civil engineer and construction manager, is remembered for her significant accomplishments in the construction of the Brooklyn Bridge. The inscription on the East Tower of the bridge reads (placed there in 1953):

> The Builders of the Bridge
> Dedicated to the Memory of
> Emily Warren Roebling
> 1843–1903
> whose faith and courage helped her stricken husband
> Col. Washington A. Roebling, C.E.
> 1837–1926
> complete the construction of this bridge
> from the plans of his father
> John A. Roebling, C.E.
> 1806–1869 who gave his life to the bridge
> BACK OF EVERY GREAT WORK WE CAN FIND
> THE SELF-SACRIFICING DEVOTION OF A
> WOMAN.

Without Emily Warren Roebling, the Brooklyn Bridge—one of the greatest engineering projects of the nineteenth century—might not have been completed on May 24, 1883. With the assistance of her brother and husband, Roebling learned engineering through the study of higher mathematics, strength of materials, stress analysis, the calculation of catenary curves, bridge specifications, and the intricacies of cable construction. Her engineering skills allowed her to become the principal assistant and inspector of the bridge as her husband, Washington Roebling, could no longer visit the site because he had "Bends" disease. She was able to discuss structural steel requirements with representatives from steel mills and assisted them with designs and shapes never before fabricated.

She said, ". . . I have more brains, common sense, and know-how generally than any two engineers civil or uncivil that I have ever met . . ." The bridge, with a span of 1,595 feet, was the largest suspension bridge in the world when it was completed and remains functional today [1, 35, 36].

Edith Judith Griswold (1863–Unknown)

Renowned as a lawyer and patent expert (this is how she is listed in the *Who's Who in Engineering* in 1925), Edith Grisworld spent four years at New York Normal College where she graduated with a license to teach in the New York Schools. However, she took a special course in electricity at the time (with her father's permission). She felt that she gained a great deal in the course and that her best work was always along electrical lines.

Her career as a mechanical draftsman began in 1884. In 1885 and 1886, she worked in D.J. Miller's office, one of the first cable railroad men, where she made drawings for and estimated costs associated with cable railroads. All of her subsequent work was in patent-office drawing. During this time, she also taught geometry and mathematics in a private school.

By 1887, she was very interested in patent law and gave up her work as a mechanical draftsman to work as a managing clerk in a patent law office and learn the profession. She attended lectures at New York University Law School. In 1897, Griswold opened her own law offices as a patent attorney. She took the bar in 1898. After 1905, her health forced her to give up regular office work.

Her engineering work was primarily in mechanics, including electrical apparatus, instruments of precision, and other intricate devices. Her legal work, which came from other patent lawyers, was always (with but one exception) patents related to articles used or worn by women [14, 37].

Kate Gleason (1865–1933)

Kate Gleason began her career in the family's Gleason Works at age 11 when her brother, Tom, died. Hearing her father lament the loss of his assistant, Kate simply showed up and took his place. And, her father did not send her back home to do "women's" things; he taught her the family business. By age 14, she was the Gleason

Works bookkeeper. She became her father's indispensable assistant. In addition to keeping the books, she traveled around the country and the world selling the company's products, and serving as the public face of Gleason Works.

In 1884, she entered Cornell University's engineering program, the first woman to so enroll. However, before her freshman year was over, she needed to return to the family business, as her father could not afford the salary of the man that had been hired to take her place. Although she significantly lamented the loss of education, she was on the road by 1888, selling machines on her first road trip. By 1890, she was the Secretary-Treasurer of The Gleason Works, and its chief sales representative, a position she held until 1913. In 1893, on doctor's orders for rest, she went to England, Scotland, France, and Germany, and came back with machine orders. This was one of the earliest efforts at international marketing for any company in the U.S. Gleason learned how to turn being a female in business into an asset. She had also learned from Susan B. Anthony, one of the leaders of the suffrage movement, that any advertising is good. In 1913, family tensions, caused in large part by her being a woman in a man's world and to a widely circulated story that credited her with being the inventor of the Gleason gear planer (the inventor was her father), led to her resigning from the company.

Kate Gleason became the first woman member of the ASME in 1914. Also in 1914, she was the first woman to be appointed receiver by a bankruptcy court. She successfully undertook the reorganization of the Ingel Machine Company of East Rochester, New York. In 1916, she was one of the first women to be elected to the Rochester Chamber of Commerce and the first woman elected to the Rochester Engineering Society. She also served as president of the First National Bank, Rochester, New York, from 1917 to 1919, while its president went off to fight in World War I.

Later, Gleason became very interested in low-cost housing and built concrete houses in the Rochester area that are still inhabited today. She was the first female member of the American Concrete Institute. Gleason served as the ASME's representative to the World Power Conference in Germany in 1930. Her estate was used to establish the Kate Gleason fund, one of whose beneficiaries was the Rochester Institute of Technology (RIT). In 1998, RIT named its College of Engineering after her. Gleason attributed her success to "a bold front, a willingness to risk more than the crowd, determination, some common sense, and plenty of hard work." [1, 3, 16, 33, 38]

Bertha Lamme (1869–1954)

Bertha Lamme went to work for Westinghouse after graduating from The Ohio State University with a degree in mechanical engineering and an emphasis in electricity. She had studied electrical engineering with her brother at Ohio State "for the fun of it" and had no plans to pursue a career after earning her degree in 1893. However, she received a surprise job offer from Westinghouse where her brother Benjamin was by then employed and worked there until she married in 1905.

Bertha Lamme worked in the East Pittsburgh plant for 12 years, where she designed motors and generators. A 1907 *Pittsburgh Dispatch* article reports on her tenure at Westinghouse saying that Lamme's work in designing dynamos and motors won her a reputation "even in that hothouse of gifted electricians and inventors. She is accounted a master of the slide rule and can untangle the most intricate problems in ohms and amperes as easily and quickly as any expert man in the shop."

In 1905, she married her supervisor and retired, as required by company policy, to become a wife and mother. Her husband, Russell S. Feicht, also an Ohio State graduate, designed the 2,000-horsepower induction motors displayed at the St. Louis World's Fair in 1904, and later retired from Westinghouse as its director of engineering. The Feicht's daughter, Florence, had well-developed mathematical abilities and went on to become a physicist for the U.S. Bureau of Mines.

The Westinghouse/Bertha Lamme Scholarships were established by the Society of Women Engineers (SWE) in 1973 in honor of Westinghouse's first woman engineer [1, 3, 14].

Nora Stanton Blatch de Forest Barney (1883–1971)

Nora Stanton de Forest Barney, granddaughter of Elizabeth Cady Stanton (one of the leaders of the suffrage movement), first distinguished herself by graduating from Cornell University with a bachelor's of civil engineering in 1905. The American Bridge Company employed her, as she was in the top five of her class and a member of Sigma Pi. She became a "squad boss" after 3 weeks of employment. In 1906, she became an assistant to Lee de Forest, inventor of the radio vacuum tube and pioneer in television. They were married in 1908 and divorced in 1912. In 1909, she joined the staff of Radley Steel Construction Company as an assistant engineer and chief drafter. From 1909 to 1917, she was also active in the New York State women's suffrage movement. Then, beginning in 1912, she was an assistant engineer for the New York Public Service Commission. She married Morgan Barney, a marine architect in 1919. Barney also served as an architect and engineering inspector for the Public Works Administration in Connecticut and Rhode Island.

Besides her broad work experiences, Barney was a prolific, and widely read, writer in her field. She was actively involved in the world peace movement and the women's rights movement. Despite her many achievements, she was granted only a junior membership status in the ASCE in 1909. Nearly 14 years after being allowed to join, she filed to have her membership status elevated to "associate member," but her application was denied. She filed an appeal, but her appeal was denied. And when she attempted to regain her junior membership status, it too was denied. In her later life she became a real estate developer. In 1944, she wrote *World Peace Through a Peoples Parliament* [14, 16].

References

1. Kass-Simon G, Farnes P (eds) (1990) Women of science: righting the record. Indiana University Press, Bloomington, IN
2. Rossiter M (1992) Women scientists in America: struggles and strategies to 1940. The Johns Hopkins University Press, Baltimore, MD
3. LeBold WK, LeBold DJ (1998) Women engineers: a historical perspective. ASEE Prism 7(7):30–32
4. Barker AM (1994) Women in engineering during World War II: a taste of victory. Rochester Inst Technol unpublished
5. An overview of the United Engineering Foundation: History. www.uefoundation.org/overview.html. Accessed 15 Dec 2000
6. Founder societies of the United Engineering Foundation. www.uefoundation.org/fndsoc.html. Accessed 15 Dec 2000
7. Historical highlights. www.idis.com/aime/history.htm. Accessed 15 Dec 2000
8. ASCE profile. www.asce.org/aboutasce/profile.html. Accessed 15 Dec 2000
9. 150 Years of civil engineering. www.asce.org/150/1506years.html. Accessed 15 Dec 2000
10. Elsie Eaves—pioneer from the west. SWE Newsletter, May 1959, p. 3
11. WAAIME. www.idis.com/aime/WAAIME.HTM. Accessed 19 Dec 2000
12. The woman's auxiliary to the American Institute of Mining, Metallurgical, and Petroleum Engineers, Inc. A Division of the Society for Mining, Metallurgy and Exploration (SME). http://www.smenet.org/waaime/. Accessed 5 Apr 2015
13. The history of ASME international. www.asme.org/history/asmehist.html. Accessed 15 Dec 2000
14. Ingels M. Petticoats and slide rules. Western Soc Eng. Accessed 4 Sept 1952, and later published in Midwest Engineer
15. The origins of IEEE. www.ieee.org/organizations/.ical_articles/history_of_ieee.html. Accessed 15 Dec 2000
16. Read PJ, Witlieb BL (1992) The book of women's firsts. Random House, New York
17. Rossiter MW (1995) Women scientists in America: before affirmative action 1940-1972. The Johns Hopkins University Press, Baltimore, MD
18. Welcome to the IEEE fellow program: our history. www.ieee.org/about/awards/fellows/fellows.htm. Accessed 19 Dec 2000
19. Goff AC (1946) Women can be engineers. Edwards Brothers, Inc., Ann Arbor, MI
20. www.swe.org/SWE/Awards/achieve3.htm. Accessed 1 Sept 1999
21. A brief history of AIChE. www.aiche.org/welcome/history.htm. Accessed 20 Aug 1999
22. www.tbp.org/TBP/INFORMATION/Info_book/membership. Accessed 20 Aug 1999
23. Tau Beta Pi: integrity and excellence in engineering. www.tbp.org. Accessed 20 Aug 1999
24. The Phi Beta Kappa Society: a short history of Phi Beta Kappa. www.pbk.org/history.htm. Accessed 19 Dec 2000
25. Society history. www.asse.org/hsoci.htm. Accessed 19 Dec 2000
26. (1948) The outlook for women in architecture and engineering. U.S. Government Printing Office, Washington, DC, p. 5–33. Bulletin of the Women's Bureau No. 223-5
27. www.iienet.org/Aboutg.htm. Accessed 20 Aug 1999, p. 1
28. About the NAE. www.nae.edu/nae/naehome.nsf/weblinks/NAEW-4NHMQM?opendocument. Accessed 19 Dec 2000
29. www.iienet.org/historg.htm. Accessed 20 Aug 1999
30. Schneider D, Schneider CF (1993) The ABC-CLIO companion to women in the workplace. ABC-CLIO, Santa Barbara, CA
31. Shearer BH, Shearer BS (eds) (1997) Notable women in the physical sciences: a biographical dictionary. Greenwood Press, Westport, CT
32. Oglivie MB (1993) Women in science: antiquity through the nineteenth century, a biographical dictionary with annotated bibliography. MIT Press, Cambridge, MA

33. (1996) Webster's Dictionary of American Women. SMITHMARK Publishers, New York
34. Richards E. www.greatwomen.org/rchrdse.htm. Accessed 26 May 1999
35. Weigold M (1984) Silent builder: Emily Warren Roebling and the Brooklyn Bridge. Associated Faculty Press, Inc., Port Washington, NY
36. (1959) Landmarks of the world: Brooklyn Bridge. Holiday, The Curtis Publishing Company, pp 125–129
37. Lee EC (ed) (1925) The biographical Cyclopaedia of American Women, vol II. The Franklin W. Lee Publishing Company, New York
38. Bartels N. The first lady of gearing. www.geartechnology.com/mag/gt-kg.htm. Accessed 2 Sept 1999

Chapter 3
War's Unintended Consequences

Abstract During World War I, women were encouraged to participate in the work force and support the war effort. Many historians believe that the vote for women's suffrage, which was finally ratified in 1920, was out of gratitude for women's efforts during the war. However, in spite of that "gratitude," most women were not allowed to keep the jobs they had filled during the war when it was over and made little headway in the 1920s and 1930s. During World War II, women were encouraged to enter the work force because the men were gone. Rosie the Riveter, the personification of women's contributions to the war effort, put a public face on the need for women to fill positions to keep war materials rolling off the assembly lines. Women were trained as engineers, engineering aides, and engineering cadettes. After the war ended, women were displaced in institutions of higher education and were no longer welcome in engineering careers either. But after this war, women reacted differently. They established the Society of Women Engineers and began to encourage each other to pursue scientific and technical careers. Profiles are provided for scientific and engineering women from the late 1800s through the late 1900s.

World War I

During World War I, women were encouraged to participate in the work force and support the war effort. This "war to end all wars" has also been called the first engineers' war, as the results of military applications of scientific and technological advances appeared as tanks, airplanes, and submarines [1]. Because of the shortage of male engineers during World War I, women had the opportunity to work in factories and offices where they were instrumental in keeping the manufacturing industry working [2]. They were drawn into tool design and chemical research, and designed buildings and automobiles [3].

However, because the U.S. was involved in the conflict for only 19 months, the manpower emphasis was on the emergency training of mechanics to service the new military technology and not the education of engineers with degrees [1]. Some women became war-time mechanics through special training; some of those women, because of this introduction, actually graduated with engineering degrees in the 1920s. Approximately 120 women, some without any college degree, worked

© The Editor(s) (if applicable) and the Author(s) 2017
J.S. Tietjen, *Engineering Women: Re-visioning Women's Scientific Achievements and Impacts*, Women in Engineering and Science,
DOI 10.1007/978-3-319-40800-2_3

during the war in engineering jobs. The end of the war signaled the end of this opportunity. However, prior to 1920, at least 45 women are known to have graduated from college with engineering degrees [4]. It is ironic that war, an activity that women tend to oppose, increases women's economic progress, as men are drawn out of the civilian workforce [3].

Many historians believe that the vote for women's suffrage, which was finally ratified in 1920, was out of gratitude for women's efforts during the war [5]. However, in spite of that "gratitude," most women were not allowed to keep the jobs they had filled during the war when it was over. The few women who did remain in engineering after World War I must have been nearly invisible, because according to the *American Men of Science* directory for 1921, there were no women at all in engineering, although the 1920 Census reported 41 women out of a total of 130,000 engineers [5, 6].

One of these 41 women was probably Elsie Eaves, a 1920 civil engineering graduate from the University of Colorado who had just embarked on what would be an illustrious career. Eaves was instrumental in establishing a national organization for women engineers: the Society of Women Engineers and Architects. In 1919, she and several of her colleagues at the University of Colorado wrote to engineering schools across the country, asking for information on women engineering students and graduates. They found 63 women enrolled in engineering at 20 universities—including 43 at the University of Michigan alone! The Michigan women had organized a group of their own, the T-Square Society in 1914. Most schools, however, did not yet admit women to their engineering programs. One professor responded to Eaves' letter: "I would state that we have not now, have never had, and do not expect to have in the near future, any women students registered in our engineering department" [7]. In the end, there were too few women engineers around the country and they were too scattered geographically to keep the organization alive, so it folded [7].

Women engineers made little headway in the 1920s and 1930s. Despite the economic boom in the 1920s and the accompanying optimism and expansion, women in the professions generally did not benefit from the strong economy. Out of 197 fellowships awarded in engineering during one period of the 1920s, one went to a woman. However, the number of women engineering graduates did grow by 113 from 1920 to 1929, reaching 158. And 53 engineering colleges (an increase from 35 in 1920) now admitted women [4].

Those women who were able to graduate with a degree and find employment still faced significant obstacles in advancing in their careers. After graduation, a man would start as a junior engineer and could generally expect, if his performance was satisfactory, to be promoted to senior engineer and then project engineer. Then, he would move on to management, with each step representing increased status, power, and salary. Women engineers, however, were usually limited to desk work, sometimes bordering on clerical work, and did not have access to plant or field work, which denied them the necessary experience to move up the ladder. Many higher level positions required travel to remote locations, behavior not acceptable for "gentle women." Women often needed advanced training—not generally required of men. Although a few women were able to hold positions with professional responsibility, many others operated on the fringes of the profession as writers, editors, secretaries, librarians, industrial teachers, or laboratory assistants—all more acceptable positions for women than being an actual engineer [1].

Great Depression

Men and women alike were forced to cope when the stock market crashed in 1929, and the Great Depression followed. Many women had to sacrifice personal ambition and accept a life of economic inactivity [2]. Thus many women engineers either voluntarily or involuntarily ceased their careers after 1929. Jobs in general were scarce, and what jobs were available were unlikely to be given to women. And prejudices for men in the engineering fields were obvious as evidenced in this advertisement for the Colorado School of Mines from the 1930 Adelphian Yearbook (Fig. 3.1) [8].

However, as college enrollments dwindled, colleges began to look for women as students. Twenty-seven more engineering schools began to admit women and the number of female engineering graduates increased by 156 during 1930–1939. With a new emphasis on graduate education for engineers as a result of technology developments during World War I, it is interesting that six schools that excluded women at the undergraduate level allowed them to pursue graduate engineering degrees. Prior to 1942, 18 women had received master's degrees in engineering across the entire U.S. One woman did receive an engineering Ph.D. in 1920, but no more

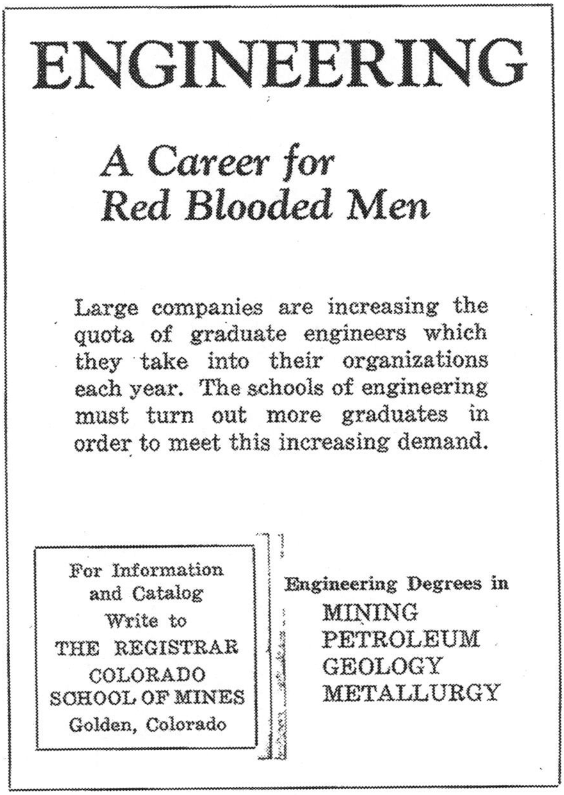

Fig. 3.1 1930s era recruitment advertisement for engineering students

ENGINEERING

A Career for Red Blooded Men

Large companies are increasing the quota of graduate engineers which they take into their organizations each year. The schools of engineering must turn out more graduates in order to meet this increasing demand.

For Information
and Catalog
Write to
THE REGISTRAR
COLORADO
SCHOOL OF MINES
Golden, Colorado

Engineering Degrees in
MINING
PETROLEUM
GEOLOGY
METALLURGY

Table 3.1 Distribution by branch of engineering of men and women employed as professional engineers, 1940

Branch of engineering	Total	Number		Percent		Percent women are of total
		Men	Women	Men	Women	
All employed professional engineers	245,288	244,558	730	100.0	100.0	0.3
Civil engineers	80,362	80,171	191	32.8	26.2	0.2
Mechanical engineers	82,443	82,255	188	33.6	25.8	0.2
Electrical engineers	53,267	53,103	164	21.7	22.5	0.3
Industrial engineers	9283	9209	74	3.8	10.1	0.8
Mining and metallurgical engineers	8813	8739	74	3.6	10.1	0.8
Chemical engineers	11,120	11,081	39	4.5	5.3	0.4

women earned that capstone degree again until three women did so in the 1930s. However, women still accounted for only about three out of every 1000 engineers. And no women were on the engineering faculty at any of the 20 largest doctoral universities in the country in 1938 [1, 4].

By 1938, the percentage of women in engineering represented significantly less than one half of one percent of all engineers. Although, it was estimated that about 1000 women engineers and architects were trained in the U.S., the 1938 edition of *American Men of Science* reported eight women, representing 0.2 % of its approximately 3500 engineers [5, 9]. The 1940 Census listed 730 women employed as engineers (see Table 3.1), less than 0.3 % of the total; but many of those are thought not to have the education credentials [10]. And in 1938, there were zero women engineers among the almost 20,000 engineers employed in the Federal classified civil service [10].

The highly visible women described in our profiles either were linked with a male relative who was making a significant impact in the engineering field, or were so outstanding on their own that their accomplishments could not be ignored. Female engineering students were still aliens in a man's world and had to deal with both implicit and explicit constraints in their educations and in their jobs, constraints which often relegated them to the margins of the profession. In addition, the number of women in engineering was so small, that no feminized branches of engineering even had a chance to develop [1]. Most women were still not admitted to the professional engineering societies in their field of expertise, and a nationwide organization for women engineers had not yet been established.

World War II

World War II, similar to World War I, presented opportunities for women to assist in the war effort. This movement of women back into the work force is often personified by famed metalworker and poster woman "Rosie the Riveter." Rosie is portrayed as a powerful woman, pictured in a headscarf while displaying large arm muscles (her sleeves are rolled up) with the caption "We Can Do It!" [2, 9] Although World War I had seen the serious use of technology for airplanes, submarines, tanks, bombs, and other equipment, technological development and organized scientific research in the ensuing 20 years was such that the machines of World War II were more destructive and more powerful. Consequently, they needed engineers and technicians to design, build, and maintain them. Many of the men with the skills to fill these engineering and technical jobs were now in the military [1].

Women were encouraged to enter the work force because the men were gone. Indeed, they began to be invited into the good positions, not just those invisible jobs discounted as "women's work" (such as silk hose manufacturing) [1]. People were needed at drawing boards; in engineering shops to keep planes, tanks, and other essential materials rolling off the assembly line—everywhere that the war effort needed support. The problem for the government was that with the men gone, women and blacks, the two reserve labor forces of the country, were the only ones available to fill these positions. But, there were not enough appropriately trained women or blacks. In engineering, the shortages were so great that women were vigorously recruited during World War II [9, 11].

The Office of War Information (OWI) and the War Manpower Commission began issuing literature and propaganda glorifying women as scientists and engineers to bolster the war effort. In 1942, a film, *Women in Defense*, narrated by Kathryn Hepburn from a script written by Eleanor Roosevelt, urged women to go to work in government or on scientific projects. In 1942, the movie *Madame Curie*, starring award-winning Greer Garson, was released, further glorifying women's contributions in science. New training programs for scientists and engineers were established and, by 1943, women (and to some extent, blacks) were being specially recruited and trained for jobs in industry. Bright high school students were sought out and urged to major in science in college through the Westinghouse National Science Talent Search (established 1943) and the Bausch and Lomb Science Talent Search (established 1944). Both programs had women among their early winners, and the Westinghouse program required that the percentage of women winners be proportional to female entrants. Books and articles were released during the 1942 through 1945 period urging women to pursue careers in science and engineering. Edna Yost's book *American Women of Science* (1943), which lauded women's past and current contributions to science, painted a bright picture for women in these fields [9].

To fill the personnel gap, the U.S. government began running crash courses in science, engineering, and management for both men and women after the war began. Some specific courses were targeted at helping women become engineering aides and engineering cadettes. Many engineering schools set up special engineering training courses for women sponsored by the War Department, the Signal Corps, the Ordinance Department, and the Air Force. Twenty-nine institutions that had heretofore excluded women including the Carnegie Institute of Technology, Columbia University's School of Engineering, and Rensselaer Polytechnic Institute, admitted women engineering students for the first time between 1940 and 1945, to train them in support of the war effort [1, 2, 9–11].

Engineering Aides, Engineering Cadettes and Engineers

In 1942 and 1943, the call went out for more women engineers. Elsie Eaves (profiled at the end of this chapter), of the *Engineering News-Record,* warned women that the word "engineer" was being applied to two very different career paths: Those few women with degrees in the field could expect to be hired into professional positions. But those non-engineering college female graduates with a few additional courses in drafting and machine testing, and those women without college degrees, were hired into subprofessional jobs as "engineering aides" or "engineering cadettes." Sometimes, they were temporary assistants to men who had been promoted from lower positions within the organization. Nonetheless, there were plenty of jobs for both kinds of "engineers" at this point. Even the federal government changed its longtime policy and began to hire women engineers in 1942 [9].

The demand was so great for the subordinate type of personnel (those classified as "aides" or "cadettes") that a special program was set up by the federal government as early as October 1940 in anticipation of entering World War II. The U.S. Office of Education administered this "defense training" program that was funded by a special appropriation. It was one of the first federal government efforts to increase and train scientific manpower. Manpower shortages were expected to be severe if the U.S. entered the war—the aircraft industry on Long Island and in New Jersey alone would require about half of all the engineering graduates nationwide. The first Emergency Defense Training (EDT) program course was offered in December 1940, and by June 1941, over 100,000 people (almost all men) had received training in engineering subjects [1, 9].

At first, the Engineering, Science, and Management War Training program (ESMWT) offered training only at 4-year technical schools. Colleges were responsible for determining local training needs and developing courses to meet those needs, but not to provide a complete engineering education nor necessarily any academic credit for the courses [1]. As the war continued and personnel shortages intensified, the program and the corresponding training was extended to a total of 227 colleges and universities, including several women's colleges, where courses were taught in elementary engineering, mathematics, chemistry, physics, and safety engineering. The ESMWT opened doors of opportunity to many women and some minorities [9]. In a report issued in December 1942, the ESMWT reported classes offered at Clarkson, Cornell, the University of Rochester, Syracuse University, Rensselaer Polytechnic Institute, and Union College. Union College had, in fact, given courses to 145 women out of 817 students, including a special course in electricity and mechanics for "girl high school graduates employed on testing work in the Engineering Department of the General Electric Company" [1].

In 1943 and 1944, 21.8 % of the enrollees were women and minorities. By 1945, over 1.8 million people had been trained including 280,000 women (15.7 %) and 25,000 blacks (1.4 %) [9]. About one-fifth of these women were enrolled in engineering drawing, while significant numbers were learning aeronautical engineering, inspection and testing, mapping and surveying, and engineering fundamentals; the rest were registered for other scientific and management courses [1].

Other departments of the federal government conducted their own programs to recruit and train women for technical positions. The Office of the Chief of Ordnance trained women with high school mathematics for civil service appointments as junior engineering aides during a 3-month intensive course at the University of Michigan. Female engineering aides for the Frankford and Picatinny Arsenals were trained at similar programs sponsored at Rutgers University, Drexel Institute, Temple University, and the University of Pennsylvania. The Signal Corps and the Army Air Forces provided technical training to hundreds of women who were often then assigned clerical duties, which resulted in a high turnover rate [1].

Some companies had other training requirements or couldn't wait for the results of the government programs to bear fruit. General Electric recruited women with degrees in math or physics and then gave them on-the-job training so that they could handle computations in GE's machine-tool department. However, so few women existed with these types of credentials, that the companies began reaching down further, pursuing women still in college [11].

Aircraft companies started special programs for engineering aides and engineering cadettes. The Goodyear Aircraft Corporation developed a 6-month program in aeronautical engineering at the University of Cincinnati to prepare women as "junior engineers" [1].

The Vega Aircraft Corporation needed aircraft design engineers starting in 1941. In cooperation with Lockheed Aircraft Corporation and the California Institute of Technology, their engineering shortage was temporarily solved by hiring engineers trained in other disciplines—civil, mechanical, or electrical—from other industries and training them for aircraft work with 8 weeks of college-level instruction at Cal Tech, followed by 8 weeks of an on-the-job apprenticeship. Other technical and clerical employees with some experience took evening classes specially arranged with the University of California in aircraft engineering-related subjects. But soon, all available candidates for these programs had been trained.

Next, Vega established a full-time training program at the University of California at Los Angeles (UCLA) to teach drafting for women. These women were not already employees, but were guaranteed jobs if they finished the course.

By 1943, Vega established a full-time 52-week program at Cal Tech for two groups of 20 full-time employees. The 40 students, all women, ranged in age from 18 to 49 years old, and already worked in the engineering department at Vega. They were able to take a full college course in engineering, stripped of all non-essentials, and earn the chance to be upgraded to higher level jobs [1].

In 1942, the Curtiss-Wright airplane company developed a plan for training "CW Cadettes," young women with at least 2 years of college education, including 1 year of college-level mathematics. Seven colleges, some of which had never previously enrolled women—Cornell, Penn State, Purdue, Minnesota, Texas, Rensselaer Polytechnic Institute, and Iowa State—gave over 600 women a 10-month crash course of engineering, math, job terminology, aircraft drawing, engineering mechanics, airplane materials, theory of flight, and aircraft production. After their training, these cadettes were assigned to airplane plants to work in research, testing, and production [1, 2, 11].

The Engineering Cadette Program in the School of Electrical Engineering at Purdue University sponsored by Radio Corporation of America (RCA) helped convince skeptical faculty that women could excel technically and academically. Cadettes accounted for 20 % of the engineering staff in some RCA plants, allowing women to move beyond being tokens [1].

By 1945, the number of women engineers listed on the National Roster had increased from 144 in 1941 to 395 (an amazing increase of almost 300 % in 4 years), mostly at the bachelor's level. And many more engineering colleges and universities now at least had seen women as students [9]. The number of women enrolled as engineering undergraduates had increased to 1800 by 1945 [1].

Faculty

Women also were encouraged to become faculty members at colleges and universities, temporarily replacing men who had been called into government service or into the military. In fact, a significant number of female scientists, including mathematicians and chemists, who were not working on other war projects, moved into the ranks of college and university faculty. In 1942, 2412 women scientists (including 50 women in engineering out of a total of 5394 engineering teachers) were on science faculties. By 1946, the number had increased to 7746 (an over 200 % increase in 4 years) and the number of women engineering faculty, including some women who taught at all-male institutions, had increased to 53 [1, 9].

These women cadettes, engineering aides, engineers in industry, and faculty all made significant contributions to the war effort, and to the advancement and acceptance of women in technical fields. All in all, about 45,000 women were trained for engineering jobs during World War II [1].

After World War II

When the war ended, millions of veterans—primarily men—came home expecting to be gainfully employed. Many of the women who had been employed found themselves no longer welcome in the work force. In fact, these women were now expected to go home and raise babies or "go back to the kitchen" and not steal work from returning GIs. The special educational programs were shut down. And because many engineering schools still did not admit women, educational opportunities for women who wanted to pursue an engineering career were once again limited. By January 1946, 4 million women had left the labor force [1, 2].

By the time World War II ended, however, women were not quite as eager to accept the inevitable as they were when World War I ended. They began generating statistics and scientific data to prove they were just as capable as men and were being treated unfairly. Still, even for those who managed to hold on to their positions, they were unable to advance as men did. Women were still channeled into less

Table 3.2 Women earning undergraduate engineering degrees 1948–1953

Year	Number of degrees
1947–1948	191
1949–1950	171
1951–1952	60
1952–1953	37

challenging work, kept in lower ranks, and paid lower salaries by employers who minimized their contributions [9].

A very significant impact of the war's end was the GI Bill, which provided funds for veterans' education. Schools that had been underutilized during the war were suddenly faced with a deluge of male students. It is estimated that 7.8 million veterans chose to take advantage of the GI Bill's educational provisions. Female enrollment at some coeducational schools had reached 50 % or higher during the war, but as the demand for slots increased, many institutions introduced or reintroduced maximum quotas on female enrollment. Some colleges refused to accept applications from out-of-state women. Women were told that space was not available and that they would have to attend other institutions. For a few years, women who were already in the educational pipeline, women veterans and wartime women "engineering aides" getting engineering undergraduate degrees, contributed to a brief expansion in the number of women getting engineering degrees. Through the 1950s, the percentage of women dropped to just 25–35 % of the total college undergraduate enrollment. The number of women graduating with engineering degrees declined as well as shown in Table 3.2 [9].

Discrimination against women was widespread; women were systematically pushed out of science and engineering at the undergraduate level, at the graduate level, and as faculty. Men even took over women's dormitories and many universities added temporary housing to accommodate the returning male GIs. Male veterans replaced female staff and faculty as well as female engineering students. There were no discrimination laws in place (these would not be enacted until the 1960s or later) to prevent any of this behavior. The nation was also extremely grateful to its veterans and felt there was, in many cases, a duty to make their adjustment to civilian life as easy as possible. Objections or protests were few and far between and not effective. And, society as a whole was still quite ambivalent about the proper role for women. These ambitious women in engineering who were certainly atypical, threatened presumed male and female spheres [9].

The *Outlook for Women in Architecture and Engineering*, a Bulletin of the Women's Bureau of the U.S. Department of Labor, painted a bleak picture for women in engineering [10]:

Advancement for women in engineering is conceded to be difficult. They seldom follow the usual line from junior engineer to senior engineer to project engineer, nor are they often transferred to nonengineering work in sales, purchasing or administration. Usually limited by custom to office work, as compared with field or plant work, women engineers rarely find opportunities to obtain the rounded experience necessary for normal progression. The fact that many of the positions representing advancement often require field work or travel to remote locations further reduces their chances. However, a few women have broken through these bounds.

Table 3.3 Membership in the founder societies, 1946 [10]

Organization	Total membership	Number of women members
American Institute of Electrical Engineers	24,526	14
American Society of Civil Engineers	21,100	23
American Society of Mechanical Engineers	20,060	33
American Institute of Mining and Metallurgical Engineers	12,600	26
American Institute of Chemical Engineers	5788	5

Women engineers, who had been welcomed with open arms during the war years, were now told to abandon the idea of an engineering career. The Women's Bureau reported that there were 950 women in engineering as of 1946–1947. These women constituted only 0.3 % of the total 317,000 engineers in the U.S. Of course, this was much higher than the eight women reported to be practicing engineering in the 1949 edition of *American Men of Science,* which despite its title, reported on all individuals practicing in the sciences. The U.S. Census of 1950 counted 6475 women employed as engineers, of whom only 41 % had 4 years or more of college, and 17 % had not graduated from high school. Probably about 3600 of these women were truly engineering professionals. There were now an adequate number of women in engineering to finally constitute enough of a critical mass to move the cause of women in engineering forward, both to seek more women as engineers and to improve career opportunities for those already in the field [1, 7, 9, 12].

In fact, women seeking careers or advancement in their existing careers and students studying engineering started to band together for mutual support and job-hunting help. In the late 1940s, the war-support effort and the associated defense industry had made cutbacks across the board. Employment for many engineering disciplines was at a low. Many professional societies still excluded women, either outright or by refusing them full membership, and thus had very few women among their membership as shown in Table 3.3. But, this time, efforts for mutual support among women engineers were enabled by a nationwide infrastructure that had not existed in 1919. Now there was the telephone, the automobile, the superhighway, and the railroad [7].

Female engineers fueled by isolation and discrimination, and the national infrastructure, founded the Society of Women Engineers (SWE) in 1949–1950 [2].

Society of Women Engineers

SWE traces its founding back to a group of female students at Drexel University in Philadelphia and a gathering of Cooper Union (an engineering college in New York City) and City College of New York graduates at Green Engineering Camp of the Cooper Union in New Jersey. The Drexel students had begun meeting in 1948, calling themselves the "Philadelphia District of the Society of Women Engineers." The Camp Green meeting, held May 25, 1950, actually resulted in the official founding of SWE. The first president, Beatrice Hicks, was elected, membership requirements were established, and dues were set. Women attending the Camp Green meeting were also from Boston and Washington, D.C. and represented two major constituencies: (1) the pioneers—those who had been trained and were working as engineers prior to or during World War II; and (2), the engineering aides who had returned to college after the war to get their engineering degrees [7, 9, 13].

The objectives of the organization were:

To inform the public of the availability of qualified women for engineering positions; to foster a favorable attitude in industry towards women engineers; and to contribute to their professional advancement.

To encourage young women with suitable aptitudes and interest to enter the engineering profession, and to guide them in their educational programs.

To encourage membership in other technical and professional engineering societies, participation in their activities, and adherence to their code of ethics [7].

One of the first orders of business for SWE was the establishment of the Achievement Award. This award, the highest tribute given by SWE, honors a woman engineer who has made outstanding contributions in a field of engineering over a significant period of time [7, 14].

1952: Maria Telkes "In recognition of her meritorious contributions to the utilization of solar energy."

1953: Elsie Gregory MacGill "In recognition of her meritorious contributions to aeronautical engineering."

1954: Edith Clarke "In recognition of her many original contributions to stability theory and circuit analysis."

1955: Margaret H. Hutchinson "In recognition of her significant contributions to the field of chemical engineering."

1956: E'lise F. Harmon "In recognition of her significant contributions to the area of component and circuit miniaturization."

1957: Rebecca H. Sparling "In recognition of her meritorious contributions to high temperature metallurgy and non-destructive testing of metals."

The organization grew rapidly, numbering 350 in 1953 and surpassing 500 by 1958. The group received decent publicity too—early activities of SWE were reported in *The New York Times* [9, 15–18].

Finally, there was an organization willing to accept women engineers, recognize them for their accomplishments, and encourage other women to be engineers.

Key Women of This Period

Some of the key women in engineering whose most significant accomplishments occurred from World War I to the early 1950s had engineering education credentials while others were educated in the sciences or other fields. These women were a very small fraction of the engineering workforce and their contributions came in spite of significant discouragement and discrimination.

Mary Engle Pennington (1872–1952)

Mary Engle Pennington was the first woman member of the American Society of Refrigerating Engineers. Her picture hangs today at its successor organization—the American Society of Heating, Refrigerating and Air-Conditioning Engineers. She later became the president of the American Institute of Refrigeration. In 1947, she was elected a fellow of the American Society of Refrigerating Engineers and a fellow of the American Association for the Advancement of Science.

Pennington completed the coursework for a bachelor's degree in chemistry, biology, and hygiene at the University of Pennsylvania, but at that time (1892), the University did not grant bachelor's degrees to women. Instead, she received a Certificate of Proficiency in biology. She continued her studies and in 1895, received a Ph.D. in chemistry from the University of Pennsylvania.

Her work in refrigeration led to her appointment as head of the Department of Agriculture's food research laboratory. As she used the name "M.E. Pennington," not everyone was aware that she was a woman. In 1916, when she had been chief of the Food Research Laboratory for a decade, a railroad vice-president on whom she called, instructed his secretary "to get rid of the woman," because he had "an appointment with Dr. Pennington, the government expert."

Pennington developed standards of milk and dairy inspection that were adopted by health boards throughout the country. Her methods of preventing spoilage of eggs, poultry, and fish were adopted by the food warehousing, packaging, transportation, and distribution industries. She has six patents associated with refrigeration and spoilage prevention methods. The standards she established for refrigeration railroad cars, which she developed by riding freight trains, remained in effect for many years and gained her worldwide recognition as a perishable food expert. Pennington received the Garvan Medal from the American Chemical Society in 1940 and was the first woman elected to the American Poultry Historical Society's Hall of Fame (1947). She has been inducted into the National Women's Hall of Fame [3, 19–21].

Lillian Moller Gilbreth (1878–1972)

Lillian Moller Gilbreth, fondly referred to as the "first lady of engineering," is best known to the general population as the mother of 12 on whom the book and movie *Cheaper by the Dozen* was based. She was a pioneer in recognizing the interrelationship between engineering and human relations. Her work in industrial engineering and time and motion studies helped encourage the development of industrial engineering curricula in engineering schools. With her husband, Frank Gilbreth, who was a pioneer in scientific management and a determined researcher, Lillian Gilbreth showed companies how to improve management techniques, and how to increase industrial efficiency and production by budgeting time and energy as well as money. Her work led eventually to career suitability tests, fatigue elimination studies, and the idea of skill transfer from one job to another (the "psychology of management").

The Gilbreths worked together in many areas. They provided scientific management consulting through their firm (Gilbreth, Inc.) that advised many companies. They wrote and researched together and authored hundreds of documents. They lectured at companies, universities, and professional societies. They conducted the Gilbreth summer schools on management topics. And, of course, they raised 12 children. Lillian Gilbreth, however, continued to work in the field for decades after Frank's untimely death in 1924 and accomplished much in the field of industrial engineering alone.

Gilbreth joined the faculty at Purdue University in 1935 as a full professor of management after having served as a lecturer and advisor for a number of years. She continued to serve as an academic advisor to women students from 1948 until shortly before her death in 1972. She became head of the Department of Personnel Relations at Newark School of Engineering in 1941 and visiting professor of management at the University of Wisconsin at Madison in 1955.

In 1921, Gilbreth was named an honorary member of the Society of Industrial Engineers at a point when they still did not admit women to membership. In 1966, she was the first woman to receive the Hoover Medal for distinguished public service by an engineer. SWE named her its first honorary member in 1950 (her membership number was one) and she was one of the organization's staunchest supporters for the rest of her life. SWE established its first scholarship in 1958 and named it the Lillian Moller Gilbreth Scholarship. In 1965, she became the first woman elected to the National Academy of Engineering. The recipient of 23 honorary degrees, Gilbreth was also the recipient of the first awarding of the Gilbreth Medal from the Society of Industrial Engineers in 1931. Gilbreth was the first engineer honored by a stamp in the Great American series of stamps by the U.S. Postal Service in 1984. She was inducted into the National Women's Hall of Fame in 1995 [2, 3, 7, 22–24].

Edith Clarke (1883–1959)

Edith Clarke had always wanted to be an engineer. However, in 1908, when she graduated from Vassar, engineering was not offered nor encouraged for women. Thus, she began 3 years of work as a teacher, then enrolled at the University of Wisconsin where she studied civil engineering for a year. She then was employed by American Telephone and Telegraph (AT&T) in New York City, where she supervised women who did computations for research engineers. She enrolled at MIT in electrical engineering and received her master's degree in electrical engineering in 1919, the first woman awarded such a degree from MIT. After graduation, she had a very difficult time securing employment as an electrical engineer. Although she wanted to work at either Westinghouse Electric or General Electric, neither company had an opening for a woman engineer. In 1920, General Electric offered Clarke a job directing calculations in the turbine engine department, a job very similar to the one she had had at AT&T.

However, since she was not being allowed to do electrical engineering work, she left GE to be an instructor at the Constantinople Women's College in Turkey.

When she returned from Turkey in 1922, GE offered her a job as an electrical engineer in the central station engineering department. At GE, she became extremely interested in the system of symmetrical components, which is a mathematical means for engineers to study and solve problems of power system losses and performance of electrical equipment. Clarke was the first woman to address the American Institute of Electrical Engineers which she did in 1926 on the topic of "Steady-State Stability in Transmission Systems." She adopted this system to three-phase components (the basis of our electricity in the United States). Clarke then wrote a textbook *Circuit Analysis of AC Power Systems, Symmetrical and Related Components* (1943) and a second volume in 1950, that was used to educate all power system engineers for many years. Based on these significant contributions, Clarke was one of the first three women fellows of AIEE.

Clarke left General Electric to become a professor of electrical engineering at the University of Texas. While there, she attracted much publicity as she was the first woman professor of electrical engineering to teach in a university in the U.S. Clarke received the SWE Achievement Award in 1954 for "her many original contributions to stability theory and circuit analysis." In 2015, she was posthumously inducted into the National Inventors Hall of Fame for her invention of a graphical calculator used in the electric utility industry [3, 22, 24, 25].

Olive Dennis (1885–1957)

Olive Dennis first studied mathematics and science at Goucher College. After several years as a teacher, she completed a degree in civil engineering from Cornell (1920), with a specialization in structural engineering. That fall, she went to work in the bridge department of the Baltimore and Ohio Railroad. However, the president of the B&O railroad had other ideas about how she could contribute to the organization. After 14 months in the bridge department, Dennis was promoted to the position of Engineer of Service for the railroad, riding the rails and figuring out ways to make the railroad more accommodating to its passengers.

During her years with the railroad, Dennis pushed for better lighting and better seating (cleaner, better fabrics, and lower, reclining seats) in the coach cars. She was an advocate for air conditioning in the cars and she designed and received a patent for an individually-operated ventilator. Dennis even designed and patented the blue colonial china provided in the dining car. At the Women's Centennial Congress in New York in 1940, she was named by Carrie Chapman Catt as one of the 100 outstanding career women in the United States [3].

Elsie Eaves (1898–1983)

Elsie Eaves was elected a Fellow member of SWE in 1980, the first year SWE elected Fellows. At that time it was said that Eaves "always encouraged women by her active example and participation." A life member of SWE, she served on the

SWE Board of Trustees and had numerous firsts to her credit. She graduated from the University of Colorado at Boulder in 1920 with a B.S. in civil engineering (with honors). In her first jobs, she was a draftsman for the U.S. Bureau of Reclamation, Denver & Rio Grande Railroad, and the Colorado State Highway Department; an instructor of engineering mathematics at her alma mater; and an office engineer for Col. Herbert S. Crocker, consulting engineer, and for Crocker & Fischer, contractors in Denver, Colorado.

Then she headed to New York City and began employment with McGraw-Hill, the publishing company. Colonel Willard T. Chevalier hired Eaves (after an editor of an undisclosed organization told her "a woman's place, if not in the home, is in the department store") and created her job as assistant on market surveys for *Engineering News-Record* in 1926. She became Director of Market Surveys for *Engineering News-Record* and *Construction Methods and Equipment* shortly thereafter. In 1932, Eaves moved to the position of Manager of Business News Department, where she directed the activities of 100 staffers throughout the U.S. and Canada.

Her career in the publishing field was a series of "firsts." In 1929, Eaves originated and compiled the first national inventory of municipal and industrial sewage disposal facilities—an analysis that she recompiled at regular intervals. A few years later, she compiled statistics on needed construction, which aided the passage of the Federal Loan-Grant legislation used to revitalize the construction industry during the 1931–1935 depression. In 1945, she organized and directed the *Engineering News-Record's* measurement of Post War Planning by the Construction Industry that was used by the Committee for Economic Development and the American Society of Civil Engineers as the official progress report of the industry. This index was unprecedented in the field of engineering analysis. Under Eaves' direction, the "Post War Planning" statistics were converted into a continuous inventory of planned construction. This has become the *Engineering News-Record's* "Backlog of Proposed Construction," an index to more than $100 billion of construction activity. Another of her unique "firsts" was defining the limits and editing the pilot issues of the *Construction Daily*, a nationwide service.

Eaves's List of Firsts and Awards Are Extensive

- First woman to be licensed as a professional engineer in New York State.
- First woman member of the American Society of Civil Engineers (ASCE) (as a corporate member in 1927).
- First woman to be a life member of the ASCE (1962, at which time there were 54 women among 48,000 members).
- First woman elected to honorary membership of the ASCE (1979); first woman to be elected Associate Member, Fellow of ASCE.
- First and for a long time, the only, woman member of the American Association of Cost Engineers (1957) as well as the first civil engineer.
- First woman to receive the Honorary Life Membership Award from the American Association of Cost Engineers (1973).

- First woman to receive the International Executive Service Corporation "Service to the Country" award.
- First woman to receive the American Association of Cost Engineer's Award of Merit (1967) [3, 22, 26, 27].

Elsie MacGill (1905–1980)

Elsie MacGill was stricken by polio while she was studying for her master's degree in aeronautical engineering at the University of Michigan. She wrote her examinations from the hospital and received her degree in 1929. This was after her successful receipt of a bachelor's degree in electrical engineering at the University of Toronto (1927). For both of her degrees, she was the first woman to receive a degree in that field at that University. After her convalescence, she worked on her doctorate degree for 2 years at MIT.

Subsequent to earning her Ph.D., she joined the Fairchild Aircraft Company as an airplane designer and performed experiments in stress analysis. Later, she served as chief aeronautical engineer for the Canadian Car and Foundry Company. One of her major projects there was to build Hurricane Fighter Planes for the British Air Ministry. These planes had precise requirements for many of each plane's 25,000 parts to allow them to be interchangeable between planes. MacGill was responsible for transforming a railway boxcar manufacturer into an aircraft factory to complete this job. Later, she engineered production of the Curtiss Helldiver for the U.S. Navy.

MacGill received the SWE Achievement Award in 1953 "in recognition of her meritorious contributions to aeronautical engineering." She was also recognized for her accomplishments in Canada where she was the first woman member of the Engineering Institute of Canada and the first woman to read a paper before it. She later became the first woman to serve as a technical advisor to the UN's International Civil Aviation Organization, where she helped draft international air-worthiness regulations for commercial aircraft. She was inducted posthumously into Canada's Aviation Hall of Fame and is profiled in the Canadian Science and Engineering Hall of Fame.

MacGill was also an outspoken advocate of equal pay for equal work before the concept became popular. In 1967, she became a member of the Royal Commission on the Status of Women in Canada [3, 28, 29].

Beatrice Hicks (1919–1979)

One of the founders of the Society of Women Engineers, Beatrice Hicks served as its first president. She was committed to the organization because of her belief that there was a great future for women in engineering. Because of her interest in mathematics, physics, chemistry and mechanical drawing in high school, she decided to become an engineer. In fact, her interest had been sparked at age 13 when her engineer father had taken her to see the Empire State Building and the George Washington Bridge, and she learned that it was engineers who built such structures.

Her parents didn't actively discourage her, although her high school classmates and some of her teachers tried to discourage her, pointing out that engineering was not a proper field for women.

After her high school graduation in 1935, she entered the Newark (New Jersey) College of Engineering. Since it was during the depression, she needed to earn money for her expenses. In 1939, she received a B.S. in chemical engineering and took a position as a research assistant at the College. In 1942, she got a job with the Western Electric Company, becoming the first woman to be employed by the firm as an engineer. She worked first in the test set design department and later in the quartz crystal department. An early award citation stated "the quality of her work became legend." She studied at night while employed and, in 1949, earned an M.S. in physics from Stevens Institute. Subsequently, she undertook further graduate work at Columbia University.

When her father died in 1946, she became vice-president and chief engineer of the company he had founded, Newark Controls Company, a firm specializing in environmental sensing devices. In 1955, she bought control of the company and became president. One of the major products of the company at that time was low-water cutoffs and other devices to protect people from their own forgetfulness, often sold through mail-order companies. Here, Hicks was also involved in the design, development, and manufacture of pressure- and gas-density controls for aircraft and missiles. In 1959, she was awarded patent 3,046,369 for a molecular density scanner or gas density switch. This type of switch is a key component in systems using artificial atmospheres. After 1967, when her husband died, she became the owner of his firm, Rodney D. Chipp & Associates, a consulting firm.

In 1952, she was named "Woman of the Year in business" by *Mademoiselle* magazine. In 1961, she was the first woman engineer appointed by the U.S. Secretary of Defense to the Defense Advisory Committee on Women in the Sciences. Hicks received SWE's Achievement Award in 1963 "In recognition of her significant contributions to the theoretical study and analysis of sensing devices under extreme environmental conditions, and her substantial achievements in international technical understanding, professional guidance, and engineering education." She was the first woman to receive an honorary doctorate from Rensselaer Polytechnic Institute (1965). She also received an honorary Sc.D. from Hobart and William Smith Colleges and from the Stevens Institute of Technology (both in 1978). In 1978, she was elected to the National Academy of Engineering, the sixth woman to be elected. In 2001, she was inducted posthumously to the National Women's Hall of Fame. In 2013, Hicks posthumously received the Advancement of Invention Award from the New Jersey Inventors Hall of Fame [24, 30–32].

References

1. Barker AM. Women in engineering during World War II: a taste of victory. Rochester Inst Technol. Accessed 21 Nov 1994, unpublished
2. LeBold WK, LeBold DJ (1998) Women engineers: a historical perspective. ASEE Prism 7(7):30–32

3. Goff AC (1946) Women can be engineers. Edwards Brothers, Inc., Ann Arbor, MI
4. Turner EM. Education of women for engineering in the United States 1885-1952. UMI Dissertation Services, Ann Arbor, MI (Dissertation, New York University, 1954)
5. Rossiter MW (1992) Women scientists in America: struggles and strategies to 1940. The Johns Hopkins University Press, Baltimore, MD
6. Schneider D, Schneider CF (1993) The ABC-CLIO companion to women in the workplace. ABC-CLIO, Santa Barbara, CA
7. Kindya MN (1990) Four decades of the society of women engineers. Society of Women Engineers, New York
8. Adelphian Yearbook (1930)
9. Rossiter MW (1995) Women scientists in America: before affirmative action 1940-1972. The Johns Hopkins University Press, Baltimore, MA
10. (1948) The outlook for women in architecture and engineering. U.S. Government Printing Office, Washington, DC, Bulletin of the Women's Bureau No. 223-5
11. Bix AS. "Engineeresses" "Invade" campus: four decades of debate over technical coeducation. Proceedings of the 1999 international symposium on technology and society—women and technology: historical, societal, and professional perspectives, New Brunswick, NJ, Accessed 29–31 July 1999, pp 195–201
12. (1954) Employment opportunities for women in professional engineering. U.S. Government Printing Office, Washington, DC, Women's Bureau Bulletin No. 254
13. The SWE story. http://societyofwomenengineers.swe.org/membership/history/2454-the-swe-story. Accessed 24 May 2015
14. Individual awards overview: achievement award. http://societyofwomenengineers.swe.org/awards/individual-awards. Accessed 24 May 2015
15. Women engineers see field widening. The New York Times. p 47. Accessed 11 Mar 1951
16. U.S. Agencies seek women engineers. The New York Times. p 19. Accessed 12 March 1951.
17. Again heads women engineers. The New York Times. p 14. Accessed 6 Aug 1951.
18. Women could fill engineering jobs: Trade Society, meeting here, told they represent untapped "Source of Qualified Talent". The New York Times. p 79. Accessed 16 Mar 1952
19. Shearer BH, Shearer BS (eds) (1997) Notable women in the physical sciences: a biographical dictionary. Greenwood Press, Westport, CT
20. Oglivie MB (1993) Women in science: antiquity through the nineteenth century, a biographical dictionary with annotated bibliography. MIT Press, Cambridge, MA
21. Read PJ, Witlieb BL (1992) The book of women's firsts. Random House, New York
22. Kass-Simon G, Farnes P (eds) (1990) Women of science: righting the record. Indiana University Press, Bloomington, IN
23. Perusek A (2000) The first lady of engineering. SWE: Mag Soc Women Eng 82–83
24. Proffitt P (ed) (1999) Notable women scientists. Gale Group, Farmington Hills, MI
25. Edith Clarke I. http://invent.org/inductees/clarke-edith/. Accessed 24 May 2015
26. (1980) SWE's first fellow members: their achievements and careers. U.S. Woman Eng 9
27. Elsie Eaves scores again: is first woman honored with ASCE Life membership. p. 6. McGraw-Hill, News-Bulletin. Accessed 15 Mar 1962
28. Elizabeth Muriel Gregory MacGill. https://www.collectionscanada.gc.ca/women/030001-1409-e.html. Accessed 24 May 2015
29. Queen of the Hurricanes. http://www.cbc.ca/history/EPISCONTENTSE1EP14CH3PA2LE.html. Accessed 24 May 2015
30. Candee MD (ed) (1957) Current biography. The H. W. Wilson Company, New York, NY
31. Beatrice Hicks recognized by New Jersey Inventors Hall of Fame. SWE Magazine, Winter 2014. http://www.nxtbook.com/nxtbooks/swe/winter14/index.php#/22. Accessed 24 May 2015
32. Stanley A (1995) Mothers and daughters of invention: notes for a revised history of technology. Rutgers University Press, New Brunswick, NJ

Chapter 4
Suburbia and Sputnik

Abstract With the start of the Cold War, and then the Korean War in 1950, American women were once again asked to contribute to the nation's defense. Young women were even *encouraged* to study science and engineering, particularly after the Russian lunch of Sputnik. Similarly to the experiences after World War II, however, women were expected to be compliant about being discarded and replaced when a national crisis had passed. After the middle of the twentieth century, with higher levels of education and training among women and the general population, such treatment began to prove unacceptable for women. But it would still be several decades, and take federal legislation and Presidential actions, before significant progress was made toward even a semblance of equal opportunity for scientific and engineering women. Profiles are provided for scientific and engineering women from the mid 1900s through the early twenty-first century.

The Korean War

With the start of the Cold War, and then the Korean War in 1950, American women were once again asked to contribute to the nation's defense. Young women were even *encouraged* to study science and engineering. In 1951, President Harry Truman was seeking a standing army of 3.5 million men and highly trained human resources at home—scientists and engineers, not only for Korea but anywhere necessary for the foreseeable future. The shortage of these resources was especially acute because of low birth rates during the 1930s and a drop in engineering enrollments after World War II because of highly publicized unemployment in engineering in 1949 and 1950 [1, 2].

A need for technical manpower was growing out of the increasingly complex machines and processes used by society. A 1951 report describing the differences between a B-47 jet bomber and earlier models demonstrates the higher levels of complexity [3].

> The B-47 jet bomber, now entering volume production, required 2 years for design, 2 more years to reach test flight stage, and 2 more years to start assembly line production. A B-47 is made up of some 72,000 parts exclusive of nuts, bolts, and rivets. The B-47 requires 40 miles of wiring compared to 10 miles for the B-29. A B-47 contains over 1500 electronic tubes.

J.S. Tietjen, *Engineering Women: Re-visioning Women's Scientific Achievements and Impacts*, Women in Engineering and Science,
DOI 10.1007/978-3-319-40800-2_4

43

The wing skin must be tapered in thickness throughout its entire length from five-eighths inch at the body joint to three-sixteenths inch at the wing tip. The first B-47 plane required 3,464,000 engineering man-hours compared to 85,000 man-hours for the first production model of the B-17.

A 1951 survey by the National Society of Professional Engineers reported that 65 % of employers canvassed would hire women engineers if they were available, 45 % had found it feasible to use them, and 23 % actually employed women at that time. Although women were being sought as engineers again— at least by some employers—the women engineers in the work force were paid less than men, and their advancement was restricted, often by official company policy [3].

The Office of Defense Mobilization, in its Defense Manpower Policy No. 8 (September 1952), published a statement advocated by Arthur Flemming, assistant to the director for manpower of ODM and a strong supporter of women. The statement read: "Throughout this document all references to scientists and engineers make no distinction between the sexes or between racial groups; it being understood that equality of opportunity to make maximum effective use of intellect and ability is a basic concept of democracy." In addition, the policy's eleventh of 12 recommendations was for employers of scientists and engineers "to reexamine their personnel policies and effect any changes necessary to assure full utilization of women and members of minority groups having scientific and engineering training." Flemming was expressing what was increasingly becoming the official view that women were needed as scientists and engineers. However, full and equal opportunity for women in the engineering field had yet to be realized as the Committee of Specialized Personnel from ODM reported on December 9, 1953 [1, 3].

For the most part, the female graduate [i.e., in engineering and the sciences] has been held down as far as advance in classification and remuneration is concerned. Such action on the part of management is totally unrealistic, and in order to promote the development of our high potential of female scientists and engineers, this unrealistic sex barrier must be broken.

The federal government's official policy throughout the 1950s was to encourage women to enter scientific and technical fields and to urge employers to hire them and utilize them fully (including the federal government itself). However, no federal incentives, such as tax credits for fuller utilization of womanpower or enforcement mechanisms were put in place [1].

The pendulum had swung again toward encouraging women to be engineers and scientists. The country needed women to be in the work force and supporting the war effort when the country is at war, but then expected them to be compliant about being discarded and replaced when the national crisis passed. After the middle of the twentieth century, with higher levels of education and training among women and the general population, such treatment began to prove unacceptable for women. But it would still be several decades before significant progress was made toward even a semblance of equal opportunity for scientific and engineering women.

Off to Suburbia

In spite of official government encouragement during the period of the Korean War and, as noted above, through the early 1950s, the number of women enrolling in college and in engineering programs fell. Women constituted less than 0.5 % of the total engineering student population, and a large number of colleges and universities did not admit women as students. Indeed, during a typical year in the 1950s, women might earn 100 bachelor's degrees in engineering, and the number of engineering Ph.D's they would earn could be counted on one's fingers [4]. However, there was some good news for women wishing to pursue an engineering degree. The Georgia Institute of Technology accepted women starting in 1953, and Clemson University opened its doors to women in 1955 [1, 5].

The dwindling numbers reflected low societal tolerance for pioneering women who were now even categorized as deviant [1]. Traditional attitudes were very slow to change. Not atypical were the comments about women in a 1952 article from the *Journal of Engineering Education* [3]:

> Women have certain inherent characteristics which stand them in good stead. For instance, they are conscientious, they know how to use their hands, they are careful about detail, and quite important, they are not adverse to trying something new. Witness, for example, their proclivity to change the furniture around in the house about every 3 days to see if they can find a more efficient arrangement. This is exactly the procedure that our research scientists use; that is, if you don't know if something will work or not, try it and see. Quite often in scientific studies the going gets pretty rough and girls, being more sensitive and nervous than boys, sometimes become emotionally disturbed by overwork and the fear of failure. These troubles, for the most part, can be solved by the strategic use of a few kind words and a little human understanding. Girls will work their hearts out for you if you handle them right, which usually requires nothing more than a sincere interest in their welfare.

Women in the U.S. in the 1950s were being pulled in two directions at the same time. The average age of marriage for American women dropped to its lowest level during the period 1945–1960. The birth rate soared, especially among the college educated (the children born in this period were called the baby boomers). Marriage took precedence over careers. In addition, a mass white exodus to suburbia began and, for the first time, college-educated, middle class women had as many children as poor women did [6]. Television and advertising glorified domesticity and the housewife, especially her role as a consumer. Yet in the increasingly consumer-based economy, more workers were required to design and produce all of the new products. Thus, there were many more economic opportunities for women in the work force [7].

In the face of declining female enrollments in science and engineering and with the projected shortages of technical manpower, in April 1956, President Eisenhower, with the urging of the Office of Defense Mobilization Director Arthur Flemming, established a National (later called President's) Committee on the Development of Scientists and Engineers (PSCE) to serve as a clearinghouse for the many nongovernmental efforts being undertaken around the country to train more scientists and engineers. Interestingly, 19 men and no women were appointed to this committee, and its vice chairman seemed particularly uninterested in recruiting women into engineering.

Not much progress was made on the subject of women in science and engineering up to the point that the committee was disbanded in December 1958. However, the vice chairman's request for a breakdown of engineering data into gender would prove beneficial later for female members of the National Science Foundation's (NSF) Divisional Committee for Scientific Personnel and Education [1].

Sputnik Is Launched

With the launching of the Russian satellite Sputnik in October 1957, Americans began to focus their anti-Communist sentiment on science and education. Scientists had begun trying to increase funding and emphasize scientific education earlier in the 1950s as the Cold War intensified. When Sputnik went up, Soviet superiority in science was made quite visible to the American public. And there was a new tone of urgency in the talk about recruiting women scientists and engineers [1, 8].

In response, President Eisenhower exhorted the American people to meet the need for thousands of new scientists, saying "this [national security] is for the American people the most critical problem of all . . . we need scientists by the thousands more than we are now presently planning to have." Further, the President requested that the NSF "develop a program for collection of needed supply, demand, employment and compensation data with respect to scientists and engineers" [8].

NSF accomplished this through its Scientific Manpower Program and its two elements, Manpower Studies and the National Register. This National Register of Scientific and Technical Personnel grew out of the National Roster of World War II and subsequent efforts aimed at Cold War preparedness. The National Register had already begun to collect data in 1954 but published little prior to 1959 when in response to the Sputnik launch, Congress increased the NSF's budget. Consequently much of the data available have significant gaps, and data on women scientists and engineers is particularly incomplete [1, 8].

The NSF designed programs to provide federal assistance to the "best and brightest" in order to produce the scientists needed for the future and to gather the necessary data. As Congress discussed science budgets and fellowship programs as part of the U.S. response to the Sputnik launch (training scientists and engineers was now a matter of national survival), an article titled "Science Talent Hunt Faces Stiff Obstacle: 'Feminine Fallout'; Officials Fear Many Federal Scholarships Will Go to Girls— Who'll Shun Careers" appeared in as prestigious a newspaper as *The Wall Street Journal*. Because up to one-third of these fellowships were expected to go to women who would marry, have children, and interrupt their careers, the author commented:

> *Hence it's inevitable that some Government money will go to train scientists who experiment only with different household detergents and mathematicians who confine their work to adding up grocery bills.*

But the author further lamented it would not be feasible to place quotas on the number of fellowships given to women as this "probably would embroil the Government in a great controversy with the many 'equal rights' advocates among the ladies" [1, 8].

The National Defense Education Act (NDEA) was finally passed by Congress in 1958. The act clearly linked higher education to national defense by declaring:

> *The Congress hereby finds and declares that the security of the Nation requires the fullest development of the mental resources and technical skills of its young men and women ...*
>
> *We must increase our efforts to identify and educate more of the talent of our Nation. This requires programs that will give assurance that no student of ability will be denied an opportunity for higher education because of financial need; will correct as rapidly as possible the existing imbalances in our educational programs which have led to an insufficient proportion of our population educated in science, mathematics, and modern foreign languages and trained in technology.*

Ten new programs were established upon enactment of the NDEA, including a federal student loan program and a new graduate fellowship program larger and broader than the one at the NSF. These fellowships would continue until 1973 [1].

However, women were still feeling a conflict between their domestic obligations and pursuing a scientific or engineering career. And now with a perceived patriotic duty—especially at a time when recruitment literature stressed that Russian women constituted about half of the combined scientific and engineering workforce in that country and 25% of the Soviet Union's engineers—articles on both sides of the issue appeared in popular magazines with such titles as "Woman's Place Is in the Lab, too," "Science for the Masses," "Bright Girls: What Place in Society?" "Plight of the Intellectual Girl," and "Female-Ism: New and Insidious" [1, 8].

The first comprehensive study describing U.S. scientific and technical manpower was published in 1964—it did not examine traits such as sex and ethnicity [8]. The NSF also funded a number of studies to identify the factors associated with the low numbers of women in science and engineering. These studies showed that myths about women not being suited for engineering due to ability, emotion, or motivation were just that—myths—and the studies recommended actions to encourage women to pursue engineering [7].

Post-Sputnik, the major cultural and legislative changes of the 1960s would set the stage for greater numbers of women engineers by the turn of the century.

1960s Activism

The women's movement experienced a dramatic rebirth in the 1960s that later translated into significantly increased professional opportunities for women. That it occurred at the same time as the civil rights movement is probably much less of a coincidence than it appears. The birth of feminism and suffrage in the 1800s had been closely aligned with the abolitionist cause. Now, the rebirth of the women's movement was closely related to the struggle for racial equality. Indeed, the militancy of college students during the 1960s mirrored some of the rebellious activism that had been effective and prominent during the suffrage movement. In the 1960s, protests were held on campuses and in the streets, and students traveled to the South on behalf of civil rights [6].

Betty Friedan's *Feminine Mystique*, published in 1963, launched an attack on suburban America and the status and roles assigned to women. Friedan meant her book as a call to action, and many women strengthened their resolve to take charge of their own lives as a result of its publication. The percentage of college-educated females who worked outside the home increased from seven percent in 1950 to 25 % in 1960 [6].

In the 1960s, corrective legislation that addressed women's historically lower status in society relative to men began to roll out, one after the other. And by 1962, 53 % of all female college graduates were employed, while 36 % of those with high school diplomas held jobs. Seventy percent of all females who had 5 or more years of higher education worked [6].

Presidential Commission on the Status of Women

The Presidential Commission on the Status of Women was convened in 1961 to investigate and suggest remedies for "prejudices and outmoded customs [that] act as barriers to the full realization of women's basic rights." Seven committees representing various facets of American life—civil and political rights, education, federal employment, private employment, home and community, social security and taxes, and protective labor legislation—were involved in the commission's work. Their final report, issued in 1963, proved that in almost every area, women were second-class citizens (remember the Federal government would not even hire women engineers until 1942). President Kennedy took two actions as a result of the work that went into the commission's report:

1. Women were to be on an equal basis with men for Civil Service promotion.
2. All executive department promotions were to be based on merit [1, 6, 9].

Equal Pay Act

After the publication of the report from this Commission, and in large part because of its findings, President Kennedy signed the Equal Pay Act, which states that ". . . no employer shall discriminate between employees on the basis of sex by paying wages for equal work, the performance of which requires equal skill, effort and responsibility, and which are performed under similar working conditions." The act was sponsored by Edith Green of Oregon, one of the most influential members of Congress. It was the first major piece of legislation addressing sexual inequality since the Nineteenth Amendment. Although there were significant exemptions included as part of the Act, the legislation was an important step forward [6, 9].

Civil Rights Act: Title VII

In 1964, a second major piece of legislation—Title VII of the Civil Rights Act—was passed to prohibit discrimination in employment on the basis of race, religion, color, national origin, and sex. The original intent of the bill was to deal with racial inequality. The amendment adding the word "sex" was proposed by the powerful chair of the House Rules Committee, Howard Smith of Virginia, in an effort to retard its passage. Smith urged Congress "to protect our spinster friends in their 'right' to a husband and family," a conniving plea that was met with roars of laughter. His apparent intent was to burden the entire law with the addition of gender and cause its defeat due to the expected ensuing controversy and ridicule. Thereafter, his strategy of adding sex was referred to as a joke. Nonetheless, the amendment to the language was retained, and the law passed. The Equal Employment Opportunity Commission was formed to enforce Title VII and found that most of its complaints were from women, not from minorities as had been expected [6, 10].

The Dawn of Affirmative Action

In September 1965, President Johnson essentially began affirmative action by signing Executive Order 11246. This order required all companies wishing to do business with the federal government to not only provide equal opportunity for all, but also to take affirmative action (defined as extra steps) to bring their hiring in line with available labor pools by race [11].

Recognition of Sex Discrimination

Two years later, in 1967, President Johnson signed Executive Order 11375 extending Executive Order 11246 to include "sex" as a protected category. This executive order now required that affirmative action be taken on behalf of women in addition to minorities (as required by Executive Order 11246) so that hiring was in line with gender proportions as well as racial proportions in the relevant labor pools [11].

By the end of the 1960s, the U.S. had successfully landed men on the moon, symbolizing American technical and scientific superiority over the Soviet Union. The women's rights and civil rights movements encouraged women and minorities to pursue all career fields—including nontraditional ones such as engineering although the number of women engineering Ph.D.s in 1968 totaled 5 nationwide (0.2 % of the total). However, by the early to mid-1970s, the Vietnam War, the energy crisis, and a widening awareness of environmental issues somewhat soured Americans on science and technology. Now, scientists were needed to help save America from themselves—so maybe women and minorities, with different ways of solving problems—could help [8, 12].

1970s Progress

The 1970s represented some watershed years in the progress of women in engineering and women in the workforce, in general. In undergraduate engineering, the 1 % barrier was broken—in 1972, 525 women received B.S. degrees, a stunning 1.2 % of the total degrees. By 1979, the percentage of women receiving undergraduate degrees had increased to 9 % of the total, master's degrees in engineering were up to 5.6 %, and Ph.D. degrees amounted to 2.2 % of the total [12].

Significant legislative advances during the decade included the 1972 passage of the Education Amendments Act (particularly Title IX), the Equal Employment Opportunity Act of 1972, and the 1972 expansion of the Equal Pay Act. All served to put the public on notice that women were to be treated equally and either intentionally or inadvertently, all served to open more professional opportunities for women, including engineering opportunities.

Education Amendments Act

Title IX of the Education Amendments Act prohibited discrimination on the basis of sex in all federally-assisted educational programs. Title IX stated in part, "No person in the United States shall, on the basis of sex, be excluded from participation in, be denied the benefits of or be subjected to discrimination under any education program or activity receiving federal financial assistance."

Title IX extended the Equal Pay Act and Title VII to educational workers and applied to admissions of females to all public undergraduate institutions, professional schools, graduate schools, and vocational schools. A very significant consequence of this act was that caps on the numbers of women students accepted into medical, law, business, and other professional schools were finally abolished [11].

Certainly women still were not at parity with men in employment or education by the end of the 1970s, yet more progress had been made for women pursuing an engineering career in this decade than in any previous decade.

Key Women of This Period

The significant engineering women of accomplishment in the last half of the 20th century were educated as engineers and scientists. Their accomplishments were recognized through national honors from professional societies, governmental organizations and others.

Maria Telkes (1900–1995)

Hungarian-born Maria Telkes came to the United States in 1925 after earning a Ph.D. in physical chemistry at the University of Budapest. She worked first for the Cleveland Clinic Biophysical Laboratory. She was motivated to find alternatives to coal-fired generation for electricity production because of her horror over coal mine disasters.

When MIT received a grant from oil magnate Godfrey L. Cabot to conduct research in solar energy conversion, Telkes was appointed a research associate by MIT (1940) to start the development of semiconductors for solar thermoelectric generators. She developed solar distillers to convert sea water into drinking water for life rafts. She produced inflatable floating solar stills, weighing one pound and producing twice their weight in drinking water directly from sea water. To store solar heat for a test structure that MIT was developing, Telkes developed the use of heat of fusion of inexpensive salt hydrates that required only one-twentieth of the volume of a water tank. The phase change materials (PCM) were new in this type of application.

As a result of this work, Telkes built the first solar-heated home in 1949 in Dover, Massachusetts. She then participated in the construction of solar-heated houses by the Curtiss Wright Company in Princeton, New Jersey and at the University of Delaware from 1972 to 1977. Later, she applied heat storage principles to use off-peak electricity for heating and cooling of buildings. Her original chemical heat storage invention at MIT grew into an entirely new technology. She holds 21 patents.

During her career of over 50 years, Telkes was involved in photovoltaics, solar cooking, solar space and water heating, solar distillation, and thermoelectricity. Telkes received the first Society of Women Engineers' (SWE) Achievement Award in 1952 "in recognition of her meritorious contributions to the utilization of solar energy." Dr. Telkes, who was referred to as the "Sun Queen," invented simple solar cookers and ovens during the 1950 and 1960s that will roast, broil, and bake food without using wood, fossil fuels, or animal dung. In 1954, the Ford Foundation granted her $45,000 to develop her solar cooker. Some of her solar energy inventions can also be used for crop drying. In 2012, she was inducted into the National Inventors Hall of Fame [13–16].

Grace Murray Hopper (1906–1992)

Admiral Grace Murray Hopper was famous for carrying "nanoseconds" around with her. These lengths of wire represented the distance light traveled in a nanosecond (one billionth of a second). She was renowned for trying to convey scientific and engineering terms clearly and coherently to non-technical people.

Hopper, also known as "Amazing Grace" and "The Grandmother of the Computer Age" helped develop languages for computers and developed the first computer compiler—software that translates English (or any other language) into the 0's and

1's that computers understand (machine language). Actually, her first compiler translated English, French, and German into machine language, but the Navy told her to stick with English because computers didn't understand French and German! Computers truly only understand numbers, but humans can translate those numbers now into any of our many languages. She was also part of the group that found the first computer "bug"—a moth that had gotten trapped in a relay in the central processor. Although Admiral Hopper loved to lay claim to the discovery of this first computer "bug" (and it has been exhibited at the Smithsonian Institution's American History Museum), the term bug had actually been in use for many years.

Hopper received the SWE Achievement Award in 1964 "in recognition of her significant contributions to the burgeoning computer industry as an engineering manager and originator of automatic programming systems." She was the first woman to attain the rank of Rear Admiral in the U.S. Navy. The destroyer Hopper was commissioned by the U.S. Navy in 1997. Hopper received the National Medal of Technology from President Bush in 1991, the first individual woman to receive the medal: "For her pioneering accomplishments in the development of computer programming languages that simplified computer technology and opened the door to a significantly larger universe of users." She was inducted into the National Women's Hall of Fame in 1994. Hopper said she believed it was always easier to ask for forgiveness than permission. "If you ask me what accomplishment I'm most proud of, the answer would be all of the young people I've trained over the years; that's more important than writing the first compiler" [14, 15, 17–20].

Mary Ross (1908–2008)

The first Native American female engineer and the first woman engineer at Lockheed Missiles and Space Company, Mary Ross contributed to space exploration efforts. One of the founding members of the Skunk Works, Ross was involved in the Apollo program, the Polaris reentry vehicle, and interplanetary space probes. Ross began her career at Lockheed in 1942 and retired in 1973. A passionate advocate for women, Ross believed in the Cherokee traditions of equal education for boys and girls.

After graduating from college with a degree in mathematics, she taught high school math and science for 9 years. She completed her master's degree in 1938, while serving as a girls' advisor at a coeducational Indian boarding school. Ross joined Lockheed in 1942 as a mathematician and received intensive on-the-job training, emergency war training, and night classes to begin her engineering career. She received her engineering license in California in 1949. Later, she worked on the Poseidon and Trident Missiles.

The recipient of numerous awards, Ross was inducted into the Silicon Valley Engineering Hall of Fame and was elected Fellow of SWE. The great-great-granddaughter of a Cherokee chief, Ross inspired generations of women engineers across the U.S. [21–23].

Yvonne Brill (1924–2013)

Aerospace consultant Yvonne Brill worked tirelessly to nominate women for awards and to boards and served as a role model for several generations of women engineers, including her daughter. Her patented hydrazine/hydrazine resistojet propulsion system (3,807,657—granted April 30, 1974) provided integrated propulsion capability for geostationary satellites and became the standard in the communication satellite industry.

Brill's career began in 1945. She left to raise three children and then returned to work in her forties. In her outstanding career, she effectively expanded space horizons. Throughout most of that career, she was the sole technical woman working on propulsion systems. Her other significant technical achievements include work on propellant management feed systems, electric propulsion, and an innovative propulsion system for the Atmosphere Explorer, which, in 1973, allowed scientists to gather extensive data of the earth's thermosphere for the first time. She also managed the development, production, and testing of the Teflon solid propellant pulsed plasma propulsion system aboard the NOVA I spacecraft launched in May 1981.

Brill became a member of the National Academy of Engineering in 1987 and was a Fellow of SWE and the American Institute of Aeronautics and Astronautics. Among her many awards were the 1986 SWE Achievement Award "for important contributions in advanced auxiliary propulsion of spacecraft and devoted service to the growing professionalism of women in engineering," the 1993 SWE Resnik Challenger Medal for expanding space horizons through innovations in rocket propulsion systems, and induction into the Women in Technology International Hall of Fame in 1999. After induction into the New Jersey Inventors Hall of Fame (first woman) (2009) and the National Inventors Hall of Fame (2010), in 2011, Brill received the nation's highest honor, the National Medal of Technology and Innovation from President Obama "For innovation in rocket propulsion systems for geosynchronous and low earth orbit communication satellites, which greatly improved the effectiveness of space propulsion systems" [14, 15, 24, 25].

Thelma Estrin (1924–2014)

Pursuing her electrical engineering education at the University of Wisconsin in the 1940s and 1950s (B.S. 1948, M.S. 1949) was not easy for Thelma Estrin. Her professors did not take her seriously and because she could not get a research assistant position, her Ph.D. (1951) took a year longer than did her husband's.

Then, she had to commute for 4 hours a day to her job in New York City from her home in Princeton, New Jersey because no other opportunities were available. Nevertheless, she persevered, with the support of her husband. They had three daughters in the 1950s.

Estrin was a pioneer in the field of biomedical engineering. She was one of the first to use computer technology to solve problems in health care and medical research. Her work combined concepts from anatomy, physiology, and neuroscience with electronic technology and electrical engineering.

Estrin designed and implemented a computer system to map the nervous system. Later, she published papers on how to map the brain with the help of computers. She helped design Israel's first computer, the WEIZAC, in 1954. Estrin served as the director of the Data Processing Laboratory at the Brain Research Institute at the University of California, Los Angeles (UCLA), being barred from employment in the School of Engineering at UCLA by nepotism rules since her husband was on the faculty there. After UCLA dropped its nepotism rules in 1980, she was able to become a professor in the computer science department of the School of Engineering and Applied Science.

Not only a pioneer in her technical field, but a pioneer among women engineers, Estrin was the first woman elected to national office in the Institute of Electrical and Electronics Engineers (IEEE) as a vice president in 1981. In the late 1970s, she was the first woman to join the board of trustees of The Aerospace Corporation. Her presence and leadership on that board encouraged many women to pursue careers in aerospace engineering. She was very active in the women's movement, encouraging women to be engineers from the 1970s forward.

Among her many awards were the 1981 SWE Achievement Award "in recognition of her outstanding contributions to the field of biomedical engineering, in particular, neurophysiological research through application of computer science". A Fellow of SWE, IEEE, and the American Association for the Advancement of Science (AAAS), she was a founding fellow of the American Institute for Medical and Biological Engineering.

Her honorary doctor of science degree from the University of Wisconsin in 1989 included the following citation: "Refusing to be daunted by prejudice, she demonstrated through the undeniable quality of her work that talent is not tied to gender. She has been a model for other women who have entered and enriched the field of engineering, including two of her daughters" [14, 15, 26–28].

Jewel Plummer Cobb (1924–)

Jewel Plummer Cobb became interested in science in ninth grade when her biology teacher put a microscope in front of her. Her parents had always encouraged her interest in education and had introduced her to the wonders of science as well as strong women, particularly African-American women.

Cobb entered the University of Michigan in 1941 but stayed for just three semesters because of the racist treatment. There was no support system for black students, the dormitories were segregated, black students were not allowed in the Pretzel Ball or Beer Parlor, and women couldn't walk in the front door of the men's union building. So she transferred to Talladega College, founded by the American Missionary Society just after the abolition of slavery, and graduated with a B.S. in biology in

1944. An M.S. and Ph.D., both in cell physiology and both from New York University, followed in 1947 and 1950.

Cobb's research over the years has primarily been related to cancer causes and treatment including extensive study of melanin—the brown or black pigment that colors skin and its ability to shield human skin from ultraviolet rays. Her melanin studies involved examination of melanoma, a form of skin cancer. As her research evolved, so did her career, as she moved up the academic ladder to become President of California State University at Fullerton in 1981. In her academic administrative positions, Cobb initiated a number of programs to encourage ethnic minorities and women to pursue careers in the sciences [26, 29, 30].

Yvonne Clark (1925–)

The first woman to receive a B.S. degree in mechanical engineering from Howard University (1951), Yvonne Clark was also the first woman to receive a masters in engineering management from Vanderbilt University in 1972. The first African-American member of the SWE, Clark became a senior member and was elected a Fellow and served on many committees within the organization. In 1998, she received the Distinguished Engineering Educator Award. Clark had a distinguished career on the faculty at Tennessee State University.

Clark was raised in Louisville, Kentucky, prevented from taking a mechanical drawing class in high school because she was a woman. The University of Louisville would not admit her, being segregated at the time, and she received a full scholarship to attend Howard University. Clark faced significant discrimination in her career, often not able to obtain employment, because she was African-American and a woman. She said "They forgot to tell me I couldn't do it" [31].

Mildred S. Dresselhaus (1930–)

The first female recipient of the National Medal of Science in the engineering category, the first woman to receive the IEEE Medal of Honor, "Carbon Queen" Dr. Mildred Dresselhaus has been on the faculty of MIT since 1967. She was the first women tenured in the School of Engineering at MIT. In August 2000, she became the Director of the Office of Science in the Department of Energy, having been nominated by President Clinton and confirmed by the U.S. Senate.

As an Institute Professor of electrical engineering and physics at MIT (the first woman to be so honored), Dresselhaus is a solid-state physicist and materials scientist whose research areas include superconductivity; the electronic and optical properties of semimetals, semiconductors, and metals; and, particularly, carbon-based materials. The citation for the 1990 National Medal of Science reads "For her studies of the electronic properties of metals and semimetals, and for her service to the

Nation in establishing a prominent place for women in physics and engineering." Her 2014 U.S. Presidential Medal of Freedom citation (the highest honor for civilians), reads for "deepening our understanding of condensed matter systems and the atomic properties of carbon—which has contributed to major advances in electronics and materials research." On presenting her the award, President Obama said "Her influence is all around us, in the cars we drive, the energy we generate, the electronic devices that power our lives."

Dresselhaus received an A.B. from Hunter College in 1951 in physics and math. She received encouragement to study physics at Hunter from her advisor Rosalyn Yalow (later a Nobel Laureate in medicine) as opposed to becoming a schoolteacher. After a year in Cambridge, England, on a Fulbright scholarship in physics, she studied first at Harvard and then completed her thesis and received her Ph.D. from the University of Chicago in 1958.

The 1977 recipient of SWE's Achievement Award "for significant contributions in teaching and research in solid state electronics and materials engineering," Dresselhaus founded the MIT Women's Forum in 1970. The objective of the forum was to support the careers of women in science and engineering at MIT. In 1999, she received the Nicholson Medal for Humanitarian Science from the American Physical Society "for being a compassionate mentor and lifelong friend to young scientists; for setting high standards as researchers, teachers and citizens; and for promoting international ties in science."

In addition to her many honorary degrees, Dresselhaus has been President of the AAAS, a member of the National Academy of Engineering and the National Academy of Sciences, and a Fellow of SWE, AAAS, IEEE, and others. She attributes her success in balancing her career and raising four children to a supportive husband [14, 26, 27, 32–38].

Y. C. L. Susan Wu (1932–)

Born in Beijing, China, Ying-Chu Lin Wu was encouraged by her mother to study mechanical engineering. After graduating in 1955 (the only woman in a class with 80 men), she found that jobs for women engineers were scarce and thus she moved to the U.S. Wu earned her masters and Ph.D. degrees from The Ohio State University and the California Institute of Technology (the first woman to earn a Ph.D. in aeronautics), respectively, and found employment as an optics engineer in California. In 1965, she joined the faculty of the University of Tennessee Space Institute. There her research focus was on magnetohydrodynamics and its application to cleaner coal-fired generation.

After 23 years at UTSI, Wu left to start her own business, ERC, Inc. With more than 800 employees and annual revenues of over $100 million, ERC today does business with the Department of Defense, NASA, and others within the defense industry. Wu stepped down as president in 2000 but remains as chairman.

Wu has received many awards and honors including SWE's Achievement Award and the Outstanding Educators of America Award. She received the Amelia Earhart Fellowship in 1958, 1959, and 1962, the only three-time recipient [29, 39].

References

1. Rossiter MW (1995) Women scientists in America: before affirmative action 1940-1972. The Johns Hopkins University Press, Baltimore, MD
2. Barker AM (1994) Women in engineering during World War II: a taste of victory. Rochester Institute of Technology, unpublished
3. (1954) Employment opportunities for women in professional engineering. U.S. Government Printing Office, Washington, DC, Women's Bureau Bulletin No. 254
4. (1997) Engineering Degrees, 1996: numbers of women, minority graduates reach all-time highs, engineers: a quarterly Bulletin on careers in the profession, engineering workforce commission of the American Association of Engineering Societies, 3(1)
5. Turner EM. Education of women for engineering in the United States 1885-1952. UMI Dissertation Services, Ann Arbor, MI, (Dissertation, New York University, 1954)
6. Harris B (1978) Beyond her sphere: women in the professions in American history. Greenwood Press, Westport, CT
7. LeBold WK, LeBold DJ (1998) Women engineers: a historical perspective. ASEE Prism 7(7):30–32
8. Lucena JC (1995) "Women in engineering" a history and politics of a struggle in the making of a statistical category. Proceedings of the 1999 international symposium on technology and society—women and technology: historical, societal, and professional perspectives, pp 185–194. New Brunswick, NJ, July 29–31, 1999 Rossiter, op.cit., p. 61
9. Read PJ, Witlieb BL (1992) The book of women's firsts. Random House, New York
10. Baer JA (1996) Women in American Law: the struggle toward equality from the new deal to the present, 2nd edn. Homes & Meier, New York, p 15
11. Tobias S (1997) Faces of feminism: an activist's reflections on the women's movement. Westview Press, Boulder, CO
12. (1998) For engineering education, 1997 outputs look like 1996. Engineers. Engineering Workforce Commission of the American Association of Engineering Societies 4(1)
13. Telkes M (1900–1995) www.asu.edu/caed/Backup/AEDlibrary/libarchives/solar/telkes.html. Accessed 25 Aug 1999
14. www.swe.org/SWE/Awards,achieve3.htm. Accessed 1 Sept 1999
15. Stanley A (1995) Mothers and daughters of invention: notes for a revised history of technology. Rutgers University Press, New Brunswick, NJ
16. Maria T. Actually the name of the article is Maria Telkes, the author is National Inventors Hall of Fame, it should read Maria Telkes, as I originally wrote it. National Inventors Hall of Fame. http://invent.org/inductee-detail/?IID=468. Accessed 25 May 2015
17. Grace H (1906–1992) www.greatwomen.org/hopper.htm. Accessed 1 Sept 1999
18. Grace Murray H. The National Medal of Technology. U.S. Department of Commerce.
19. Billings CW (1989) Grace Hopper: Navy Admiral and Computer Pioneer. Enslow Publishers, Inc., Hillside, NJ
20. Zuckerman L. Think tank: if there's a bug in the etymology, you may never get it out. The New York Times. Accessed 22 Apr 2000
21. Ross MG (2008) Obituary, San Jose Mercury news. http://www.legacy.com/obituaries/mercurynews/obituary.aspx?pid=109118876. Accessed 6 June 2015

22. Ross M (2011) First native American engineer, spirited woman blogger team. http://www. thespiritedwoman.com/go_blog_blog_blog/2011/03/mary-ross-first-native-american-- engineer. Accessed 6 June 2015

23. Sheppard LM. Portfolio: profile & biographies: an interview with Mary Ross, Lash Publications International. http://www.nn.net/lash/maryross.htm. Accessed 6 June 2015

24. Yvonne Brill. www.witi.org/center/witimuseum/halloffame/1999/ybrill.shtml. Accessed 14 Feb 2001

25. President Obama Honors Nation's Top Scientists and Innovators. https://www.whitehouse. gov/the-press-office/2011/09/27/president. Accessed 25 May 2015

26. Ambrose S, Dunkle K, Lazarus B, Nair I, Harkus D (1997) Journeys of women in science and engineering: no universal constants. Temple University Press, Philadelphia

27. (1995) Who's who in technology. 7th ed., Gale Research, Inc., New York

28. Dr. Thelma Estrin. www.witi.org/center/witimuseum/halloffame/1999/testrin.shtml. Accessed 14 Feb 2001

29. Proffitt P (ed) (1999) Notable Women Scientists. Gale Group, Farmington Hills, MI

30. Shearer BF, Shearer BS (eds) (1996) Notable women in the life sciences. Greenwood Press, Westport, CT

31. SWE Magazine. Winter 2007. http://ethw.org/Yvonne_Clark. Accessed 22 Aug 2015

32. (2015) Briefings: medal of honor goes to Dresselhaus. The Institute

33. Anderson M (2015) The queen of carbon. The Institute

34. Award recipient. www.interact.nsf.gov/MOS/Histrec.nsf/. Accessed 14 Feb 2001

35. 1999 Nicholson Medal for Humanitarian Service to Mildred S. Dresselhaus MIT. www.aps. org/praw/nicholso/99wind.html. Accessed 14 Feb 2001

36. Mildred S. Dresselhaus–1997 AAAS President. www.aaas.org/communications/inside17.htm. Accessed 14 Feb 2001

37. Mildred Spiewak Dresselhaus. www.witi.org/center/witimuseum/halloffame/1998/mdressel- hau.shtml. Accessed 14 Feb 2001

38. Ortiz SJ. View from the inside: meet mildred Dresselhaus: New Director of the Office of Science. www.pnl.gov/energyscience/08-00/inside.htm. Accessed 14 Feb 2001

39. http://www.businessalabama.com/Business-Alabama/August-2013/A-Refugees-Self- Transformation/. Accessed 22 Aug 2015

Chapter 5
Bridges to the Future

Abstract The last quarter of the twentieth century saw heightened attention regarding the need for a trained and diverse scientific and engineering workforce. Several initiatives were launched but the statistics still gave credence to the phrase "Why so Few?" referring to the still low numbers of women and minorities pursuing scientific, technical, engineering and mathematical (STEM) careers. So many opportunities—and challenges—present themselves to the STEM workforce in the twenty-first century. The engineering advances in the last century are staggering from electrification to agricultural mechanization, to the internet to high-performance materials. The opportunities in nanotechnology, biotechnology and other fields as well as the challenges of aging infrastructure, terrorism, and an aging workforce promise fascinating careers as we bridge to the future. Profiles of engineering women, many of whom are still active in their careers today, are provided.

Moving Forward in the 1980s

In the early 1980s, Americans began worrying about whether or not the country was on an equal footing with technologically advanced Japan. As a result of this international competitiveness, more focus was placed on engineers and technology in the U.S., in the hopes of keeping America economically in-line. The pipeline for engineers and scientists began to be discussed, and women in engineering received significant focus and recognition [1].

The Science and Technology Equal Opportunity Act of 1980 was passed to include women and minorities as problem solvers to deal with the now recognized issues of environment, food shortages, and areas affected by affirmative action. The act said:

> . . . it is the policy of the United States to encourage men and women, equally, of all ethnic, racial, and economic backgrounds to acquire skills in science, engineering and mathematics, to have equal opportunity in education, training, and employment in scientific and engineering fields, and thereby to promote scientific and engineering literacy and the full use of the human resources of the Nation in science and engineering [1, 2].

J.S. Tietjen, *Engineering Women: Re-visioning Women's Scientific Achievements and Impacts*, Women in Engineering and Science,
DOI 10.1007/978-3-319-40800-2_5

Yet in spite of this rhetoric, the Reagan administration cut science education funding in the early 1980s. The falloff in the rate of increase in the number of women in engineering is evidenced by the flattening of the women graduates through the 1980s, and is attributed, in part, to the cuts in federal funding and the Reagan's administration significantly reduced emphasis on affirmative action [3].

Congress became convinced by 1987 that, based on manpower projections for scientists and engineers that showed significant shortfalls by 2006, something needed to be done. A law was passed creating a Task Force on Women, Minorities and the Handicapped in Science and Engineering to examine the current status of those groups in the targeted fields and to coordinate existing federal programs to promote their education and employment in science and engineering. The Task Force Report was issued in 1989 and thereafter many more educational institutions established Women in Engineering Programs and Minority Engineering Programs with the resulting available federal funding. The task force also reported that non-traditional engineers and scientists faced barriers in both promotion and progression in their careers [1, 4].

The latest national imperative to get women and minorities into technology was reflected in a 1988 report [1]:

> If compelled to single out one determinant of US competitiveness in the era of the global, technology-based economy, we would have to choose education, for in the end people are the ultimate asset in global competition. . . . But an especially important further step will be to extend the pool from which the pipeline draws by bringing into it more women, more racial minorities, and more of those who have not participated because of economic, social, and educational disadvantage. . . Thus not only is providing a better grounding in math and science for all citizens a matter of making good on the American promise of equal opportunity. It is a pragmatic necessity if we are to maintain our economic competitiveness.

Additional legislative and regulatory actions in the 1980s and 1990s helped improve the workforce for women. Specifically, the Equal Employment Commission issued regulations in 1980 that defined sexual harassment as a form of sex discrimination, thus prohibited under the Civil Rights Act of 1964. U.S. Supreme Court rulings in the 1980s and 1990s further clarified the situations constituting sexual harassment [5]. Women engineers, who tended to be fairly isolated in work environments because of their low numbers, now had legal recourse for some of the more blatant behavior they were experiencing.

The 1990s

In the 1990s, the U.S. was focused on ways to remain globally competitive with the entire world, not just Japan [1]. Additional initiatives, task forces, studies, and conferences occurred to further examine what often boils down to the phrase "Why so Few?" Why aren't more women pursuing engineering careers? And, although many issues have been identified and solutions proposed, the number of women in engineering failed to increase significantly during the 1990s [3, 6].

National Academy of Engineering

The National Academy of Engineering (NAE) launched an initiative in 1997 to examine and take positive steps on a national scale toward resolving the issue of why so few women are entering the engineering field. A web site was set up and a major conference, the Summit of Women in Engineering, was held in May 1999 [7]. One of the legacies of that effort is the web site engineer girl [8].

The Commission on the Advancement of Women and Minorities in Science, Engineering and Technology

The Commission on the Advancement of Women and Minorities in Science, Engineering and Technology (CAWMSET), established by Congress in 1998, again examined the issues and potential remedies associated with the low participation of women, minorities, and persons with disabilities in science, engineering, and technology careers. The Commission's Report *Land of Plenty* (September 2000) identified issues and made recommendations with regard to pre-college education, access to higher education, professional life, public image, and nationwide accountability [9].

Establishment of Women in Engineering Programs

By 1990, many efforts had been underway by many groups—some for more than 30 years—to increase the number of women in engineering. Universities had established Women in Engineering Programs (or Women in Science and Engineering Programs) to both recruit more women students into the field and to retain higher percentages of the female students that enrolled.

Minority Women in Engineering

The national effort on increasing the talent pool in engineering has focused on minorities as well as women for many years. As early as 1974, the National Action Council for Minorities in Engineering, Inc. (NACME) was established to lead national efforts to increase access to careers in engineering and other science-based disciplines for minorities. In 1979, the National Association of Minority Engineering Program Administrators (NAMEPA) was established to promote collaboration and cooperation among the many groups committed to improving the recruitment and retention of African Americans, Hispanics, and Native Americans in the industry. One significant initiative, announced in September 1999, the Gates Millenium Scholars Program, is an activity of the Bill and Melinda Gates Foundation.

This 20-year program will provide financial assistance to high-achieving minority students who pursue undergraduate and graduate degrees in technical fields including engineering [10–12].

Trends for the Future of Engineering

Against a backdrop of stagnant percentages of women and minorities pursuing an engineering education and a career in the engineering fields, our world has become increasingly characterized by technology. The engineering accomplishments of the 20th century have resulted in the way of living that we in the developed countries experience and expect today. The National Academy of Engineering identified the 20 top engineering achievements of the twentieth century (shown in Table 5.1): [13]

Over the history of engineering, the changes to the engineering profession and engineering education came after changes in technology and society. New disciplines were established and curricula were developed to provide the workforce necessary to support those changes [14]. It is not clear that this model should be sustainable going forward.

These advances of the twentieth century have lengthened our life spans (average life expectancy increased from 47 years of age in 1900 to 77 years of age in 2000 [15]), expanded our communications abilities by orders of magnitude, and shortened product development cycles leading to increased innovation and functionality. The world changed more in the 20th century than it had in all of the preceding years

Table 5.1 Top engineering achievements of the twentieth century

1. Electrification
2. Automobile
3. Airplane
4. Water supply and distribution
5. Electronics
6. Radio and television
7. Agricultural mechanization
8. Computers
9. Telephone
10. Air conditioning and refrigeration
11. Highways
12. Spacecraft
13. Internet
14. Imaging
15. Household appliances
16. Health technologies
17. Petroleum and petrochemical technologies
18. Laser and fiber optics
19. Nuclear technologies
20. High-performance materials

[14]. On the horizon are new breakthroughs in many fields including biotechnology, nanotechnology, computing and logistics [14]. Challenges exist with the threat of terrorism, deteriorating infrastructure, environmental concerns and meeting the needs of worldwide population growth [14].

Although projecting the future is always fraught with peril—often resulting in shattered crystal balls—we are on a path where some projections can be made about the opportunities and challenges awaiting engineers over the next decade.

Engineering Opportunities

The promises for engineering advances in the twenty-first century are numerous and exciting.

Biotechnology

Humankind is now able to attack diseases and disorders at the cellular and DNA levels leading to the dream that diseases may be eradicated and the limitations of the human body (e.g., aging) could be compensated for [14]. Tissue engineering advances and regenerative medicine provide avenues for treatment of burn victims and individuals with spinal cord injuries [14].

Biotechnology in concert with advances in nanotechnology and micro-electronic mechanical systems (MEMS) could lead to the use of tiny robots for medical treatments such as repairing tissue tears and cleaning clogged arteries. Such robots might be used to destroy cancers or change cell structures for those with inherited genetic diseases. In conjunction with advances in computer technology in the bio-technology field (bioinformatics), these types of developments could lead to individually customized drug treatments [14].

We have already seen pacemaker development, the creation of artificial organs, prosthetic devices, laser eye surgery, imaging systems, and fiber-optic-assisted non-invasive surgical techniques [14]. We have devices that monitor our health, count our steps, and show us our sleep patterns—so we can learn to behave in more healthful ways. Engineers participate in tissue engineering, drug delivery engineering, and bio-inspired computing. These fields will continue to advance and increase in scope and complexity [14].

Nanotechnology

The advances in technology at the molecular level will expand in the next decade and beyond [14]. Multiple fields will be involved—bioengineering, materials sciences, and electronics, among others. Applications for nanoengineering cover a broad range from flame retardant additives to paint pigments to biosensors as shown in Table 5.2 [14].

Table 5.2 Nanotechnology applications

Pigments in paints
Cutting tools and wear resistant coatings
Pharmaceuticals and drugs
Nanoscale particles and thin films in electronic devices
Jewelry, optical, and semiconductor wafer polishing
Biosensors, transducers, and detectors
Functional designer fluids
Propellants, nozzles, and valves
Flame retardant additives
Drug delivery, biomagnetic separation, and wound healing
Nano-optical, nanoelectronics, and nanopower sources
High-end flexible displays
Nano-bio materials as artificial organs
MEMS-based devices
Faster switches and ultra-sensitive sensors

Materials Science and Photonics

New materials and smart materials will be developed across many engineering disciplines from civil to mechanical to electrical. Civil engineers might incorporate smart materials and structures that can sense and respond such as for displacements from earthquakes or explosions. With the assumption that fuel cells will replace the existing internal combustion engines, greater knowledge and understanding of fuel-cell chemistry and materials will be required. The decrease in the size of optical devices as their power and reliability increase will drive development of photonics-based technologies. Engineers will also be required with expertise in fiber optics communications; visioning, sensing, and precision cutting for precision manufacturing applications; laser guidance; and optical sensing and monitoring [14].

Information and Communications Technology

The late twentieth century saw an explosion in communications devices that many of us now can't imagine our lives without—computers, smartphones, copiers, and the Internet. Huge volumes of information can be transmitted globally today; that capability will only increase—with the attendant responsibilities of data integrity and security. This information explosion requires expertise in materials, electronics, electromagnetics, photonics, and the underlying mathematics [14].

Logistics

The term just-in-time manufacturing hardly does justice to the intricate ballet that it necessitates [p. 16]. Worldwide coordination is required in our ever increasingly globalized world. From the tablets used by personnel in airport or to present wine lists at restaurants, wireless and inventory trading and database software have led to significantly improved productivity [14].

Wal-Mart's distribution network has been held up as the model for ensuring emergency supplies after a hurricane lands and causes widespread destruction [16]. How to most effectively and efficiently provide and move goods and services will be a focus of engineering attention in the decades to come [14].

Engineering Challenges

Just as there are opportunities, there are also challenges.

Infrastructure

The infrastructure in the U.S. is aging. Since the early 2000s, the American Society of Civil Engineers has periodically issued its report card providing grades for a wide variety of infrastructure categories. The 2013 grades, shown in Table 5.3, are not grades that anyone would have wanted to come home with or to have shown her parents! [17]

Table 5.3 2013
Infrastructure report card

Category	Grade
Water & environment	
Dams	D
Drinking water	D
Hazardous waste	D
Levees	D–
Solid waste	B–
Wastewater	D
Transportation	
Aviation	D
Bridges	C+
Inland waterways	D–
Ports	C
Rail	C+
Roads	D
Transit	D
Public facilities	
Public parks & recreation	C–
Schools	D
Energy	
Energy	D+

In some cases, the grades have decreased from report card to report card, demonstrating the deterioration of our transportation, energy, water and other facilities. An estimated \$3.6 trillion is needed through 2020 to address these concerns; much engineering expertise is required within those dollars [17].

Information and Communications Infrastructure

Although newer than the facilities referenced above, the information and communication infrastructure has shown itself vulnerable to intentional attacks, system overloads, and natural disasters. Yet, we are becoming increasingly dependent on them for our work, our banking and other financial transactions, our commerce, and our economy in general. Significant attention will be required to ensure their security and reliability [14].

Environment

The first Earth Day (April 22, 1970) came about because of a confluence of many factors leading to the realization that humankind needed to be better stewards of our planet and all it had to offer us. Population growth and increased standards of living lead to higher use of many resources as well as water. Engineers will be required to develop sustainable practices across many fields including agriculture and energy [14].

The challenges—and the opportunities—for engineers to contribute in the twenty-first century are many. New areas of inquiry have been opened up and are being pursued. The possibilities are almost endless in the ways in which engineers will help advance society and provide value in our future.

Key Women of This Period

The engineering women whose accomplishments occurred during the late twentieth century and early twenty-first century were primarily educated as engineers. Their honors and recognitions are tributes to their lasting legacies.

Ada Pressman (1927–2003)

The first three-pinner in the Society of Women Engineers (Past President, Fellow and Achievement Award), Ada Pressman was a pioneer in combustion control and burner management for supercritical power plants including the input logic and fuel air mixes associated therewith. She was directly involved in early design efforts toward more automated controls of equipment and systems, the new packaging techniques, and breakthroughs in improved precision and reliability of sensors and controls. As she progressed through the management ranks at Bechtel (earning her MBA during the process), she was recognized as one of the nation's outstanding experts in power plant controls and process instrumentation and worked on fossil-fired and nuclear power plants. Pressman is credited with significantly improving the safety of both coal-fired and nuclear power plants for both workers and nearby residents. In addition, she successfully lobbied the state engineering board in California to recognize control systems engineering as a distinct engineering field.

Pressman characterized her professional experience as including the engineering management of millions of individual hours of power generation plant design and construction and of economic studies and proposals for potential projects. She continually monitored the costs for each project as well as the technical engineering details as the design progressed. Pressman received SWE's Achievement Award in 1976 "For her significant contribution in the field of power control systems engineering."

Planning to become a secretary after she graduated from high school in Ohio, Pressman was encouraged to attend college by her father. She earned her B.S. in mechanical engineering from The Ohio State University. Pressman actively mentored women throughout their engineering careers and was devoted to promoting women's STEM careers. From her outstanding career in industry, to her many years of dedicated service to SWE, to her dedication to mentoring young women and funding their educational dreams, Pressman was truly a role model for us all [18, 19].

Sheila Widnall (1938–)

The first woman placed in charge of a branch of the military, Dr. Sheila Widnall became Secretary of the Air Force in 1993, after she was appointed by President Clinton. She was praised by the president "as a woman of high achievement, a respected scientist, a skilled administrator, and a dedicated citizen." Prior to her service as Secretary, she spent 28 years at MIT, where she had won international acclaim for her work in fluid dynamics. After her resignation from the Air Force, she rejoined the MIT faculty.

Widnall's father fostered her interest in science and math, and her working mother showed her that women can manage a career and a family. She was encouraged to pursue an engineering education. But, when she enrolled at MIT, she was one of 23 women out of a total of 936 freshmen. As she had never been part of a minority group, she experienced a culture shock. Despite this initial setback, Widnall received her B.S., M.S., and Ph.D. degrees from MIT in aeronautics and astronautics. Her first child was born 6 months before she finished her Ph.D. and her second, 4 years later. She credits a supportive husband and the ability to find good daytime child care (graduate students' wives) as contributors to her career success. She was the first MIT alumna to become a member of the faculty of the School of Engineering.

And for many years after she was hired, she was the only woman engineer on faculty. In addition, she was the first woman to head the entire MIT faculty.

Widnall is known internationally for her fluid dynamics work involving aircraft turbulence and spiraling airflows. She is the holder of three patents, has a long history of professional activities, and has received many awards. She is a member of the National Academy of Engineering and has received its Distinguished Service Award (1993). In 1998 she received the IEEE Centennial Medal. Widnall received the SWE Achievement Award in 1975 "in recognition of her significant contributions to the fluid mechanics of low speed aircraft and hydrofoils." She has been inducted into the National Women's Hall of Fame. She has served as President of the American Association for the Advancement of Science (AAAS) and as a trustee for The Aerospace Corporation.

Widnall says, "I believe that women should pursue their interest in science and engineering. The future has a way of taking care of itself if one has the proper education that supports one's dreams" [20–27].

Mary-Dell Chilton (1939–)

In 1983, Mary-Dell Chilton led the research team that produced the first transgenic plants. As such, she is considered one of the founders of modern plant biotechnology and the field of genetic engineering in agriculture. After groundbreaking efforts at the University of Washington and Washington University, she established one of the world's leading industrial biotechnology agricultural programs at Ciba-Geigy (today Syngenta). Her team has worked to produce crops with higher yields, and resistance to pests, disease and adverse environmental conditions (such as drought).

The recipient of numerous awards including the 1985 Rank Prize in Nutrition and the 2013 World Food Prize, Chilton was inducted into the National Inventors Hall of Fame in 2015. Today, Distinguished Science Fellow Chilton works in a building in the Research Triangle Park in North Carolina that bears her name.

Dr. Chilton's B.S. and Ph.D. degrees are in chemistry from the University of Illinois Urbana-Champaign. She said "My career in biotechnology has been an exciting journey and I am amazed to see the progress we have made over the years. My hope is through discoveries like mine and the discoveries to follow, we will be able to provide a brighter and better future for the generations that follow us" [28–30].

Eleanor Baum (c. 1940–)

The first female dean of any engineering college in the U.S. and the first female president of the American Society for Engineering Education, Dr. Baum is a Fellow of ABET, IEEE, and SWE. She has also served as the President of ABET.

Baum's route to engineering was not an easy one. Her high school guidance counselor thought she should study something else. In fact almost anything else would do. Her mother was very worried that people would think she was strange and, as a result, that no one would marry her. Baum was not accepted at several engineering colleges where she applied for admission because she was female; in at least one case her application was denied due to a lack of women's bathrooms. In the end, she was the only female in her engineering class at City College of New York. She received a B.S. in electrical engineering (1959) and completed her M.S. (1961) and Ph.D. (1964) degrees from the Polytechnic Institute of New York.

Baum is a national leader in engineering education and the advancement of women in science and technology. She was the 1988 recipient of the Emily Roebling Award presented by the National Women's Hall of Fame. She was inducted into the Women in Technology International Hall of Fame in 1996 and into the National Women's Hall of Fame in 2007. A recipient of SWE's Upward Mobility Award, Baum serves on the Boards of Directors of several corporations [24, 27, 31–34].

Donna Shirley (1941–)

The President of Managing Creativity and former Assistant Dean at the University of Oklahoma, Donna Shirley burst into the national and international scene in July 1997 when the Mars Pathfinder and its Sojourner Rover—the solar-powered, self-guided, microwave-oven-sized explorer, began their exploration of the Martian surface.

Shirley decided at age 10 to be an aeronautical engineer and to build airplanes. At 15, she began flying lessons and at 16, she soloed. However, her path to aeronautical engineering was not as straight as she might have envisioned.

When Shirley arrived at the University of Oklahoma in 1958 determined to study aeronautical engineering, her first visit with her adviser began with him telling her that "Girls can't be engineers." The school newspaper even ran an article noting the rarity of female engineering students—all six of them. Shirley did eventually graduate in 1962, but with a degree in journalism.

After a stint as a technical writer at McDonnell Aircraft, Shirley reapplied to the University of Oklahoma, took a leave of absence from McDonnell, and went back to engineering school. In the spring of 1965, she did graduate with a B.S. in aerospace/mechanical engineering. And she returned to McDonnell Aircraft. In 1966, Shirley took a job with the Jet Propulsion Laboratory with one objective: to get to Mars—a goal that would take 31 years.

Inducted into the Women in Technology International Hall of Fame, Shirley has written several books, including her autobiography and a book on using the collective creativity of groups to develop ideas and turn those ideas into real products [35–37].

Gail de Planque (1944–2010)

The first woman and the first health physicist to be appointed to the U.S. Nuclear Regulatory Commission, Dr. Gail de Planque was a trailblazer for women throughout her entire career. When she joined the Atomic Energy Commission's Health and Safety Laboratory (HSL) as an entry-level physicist, she was told not to expect much in the way of opportunities for advancement because women would eventually leave for marriage. She did not leave and eventually became the lab's director. During her tenure at HSL, she earned her M.S. in physics and her Ph.D. in environmental health sciences. Her master's thesis was titled "Radiation Induced Breast Cancer from Mammography"; ironically, she was later a breast cancer survivor.

Dr. de Planque was the recipient of numerous awards for her pioneering role as a woman in science and contributions to the peaceful uses of nuclear energy. One of the most significant was election to the National Academy of Engineering (NAE) with the citation "For leadership of the national nuclear programs and contributions to radiation protection devices and standards." In 2003, she received the Henry DeWolf Smith award for Nuclear Statesmanship from the American Nuclear Society and the Nuclear Energy Institute for her contributions to the peaceful use of nuclear energy. In 2015, she was inducted into the Maryland Women's Hall of Fame. Her areas of expertise included nuclear physics and environmental radiation studies.

While at the U.S. NRC, Dr. de Planque often had a pivotal role in matters relating to equal employment opportunities, flexiplace and flexitime, sexual harassment policy, and management. After her tenure at the U.S. NRC was complete, Dr. de Planque was sought after nationally and internationally including by the United Nations International Atomic Agency.

As Chair of the NAE's Celebration of Women in Engineering Steering Committee, she led the national effort to change the national dialogue on increasing the number and percentage of women in engineering and implement national and local programs that would implement new, wide-reaching efforts to get closer to parity [38].

Shirley Jackson (1946–)

The first African-American woman to receive a Ph.D. from the Massachusetts Institute of Technology (MIT), theoretical physicist Shirley Jackson is now the President of Rensselaer Polytechnic Institute (RPI). As a physicist, Jackson's area of expertise is particle physics—the branch of physics that predicts the existence of subatomic particles and the forces that bind them together.

Jackson was encouraged in her interest in science by her father who helped her with projects for her science classes. She took accelerated math and science classes in high school and graduated as valedictorian. At MIT, she was one of less than twenty African-American students on campus, the only African American studying physics, and one of about 43 women in the freshmen class of 900. After obtaining her B.S. at MIT, she opted to stay for her doctoral work in order to encourage more

African-American students to attend the institution. She completed her dissertation and obtained her Ph.D. in 1973.

After postdoctoral work at prestigious laboratories in the U.S. and abroad, Jackson joined the Theoretical Physics Research Department at AT&T Bell Laboratories in 1976. She served on the faculty at Rutgers University from 1991 to 1995 and then became the first woman and African-American Chairman of the U.S. Nuclear Regulatory Commission. In 1999, she became the first African American and first woman President of RPI.

Her numerous honors and awards include induction into the National Women's Hall of Fame and the Women in Technology International Hall of Fame, the Thomas Alva Edison Science Award, and the CIBA-GEIGY Exceptional Black Scientist. Jackson actively promotes women in science [24, 39–41].

F. Suzanne Jenniches (1948–)

Suzanne Jenniches, retired Vice President, Communications Systems, Northrop Grumman, had not even heard of the word engineering until she was 23 years old and teaching high school biology. She was influenced by the first Earth Day (April 1970) to enroll in a master's program in environmental engineering at Johns Hopkins University.

In 1978, she completed her M.S. in environmental engineering, although the large majority of her courses were undergraduate courses in computers and electrical engineering. In the interim, Jenniches had left her public education career and begun her employment with Westinghouse Electric Corporation as a product evaluation engineer.

Jenniches rose through the ranks at Westinghouse, at one point becoming the Operations Manager for B-1B Offensive Radar and Special Access Systems. The production of this radar system was the critical path for the B-1B program and was briefed monthly to then-Secretary of Defense, Caspar Weinberger. Jenniches successfully introduced the first Electronically Scanned Antenna into a production aircraft with the B-1B.

Subsequently, she became the General Manager, Automation and Information Systems for Westinghouse and later Northrop Grumman (after its purchase of this division of Westinghouse). While in this position, Jenniches oversaw the delivery of over 10,000 postal systems for the U.S. Postal Service. In addition, she led the team that designed and deployed the Federal Express Small Package Sort System in its 500,000 square-foot facility in Memphis, Tennessee. At the time of her retirement, Jenniches was leading Northrop Grumman's efforts in the Electronics Systems International Business.

Jenniches received the 2000 Achievement Award from SWE "in recognition of outstanding leadership in manufacturing innovation and for setting the highest standards of excellence in producibility engineering." She has served as the National President of SWE. Jenniches serves on corporate boards and has received gubernatorial appointments to several Maryland commissions and task forces. A Fellow of SWE, Jenniches served for many years as a consultant to the Service and Technology Board of the U.S. Army. In 2015, she received the Kate Gleason Award from the American Society of Mechanical Engineers [42, 43].

Judith Resnik (1949–1986)

A member of the ill-fated Challenger mission in 1986, Judith Resnik was a "can do" kind of person. She received her B.S. in electrical engineering from Carnegie-Mellon in 1970 and a Ph.D., also in electrical engineering, from the University of Maryland, College Park in 1977.

Resnik was selected for the astronaut corps in 1978, having previously served as a biomedical engineer and staff fellow in the laboratory of neurophysiology at the National Institutes of Health. The second American woman to travel in space, Resnik was a mission specialist on space shuttle Discovery's maiden voyage in 1984. During 96 orbits of the earth, the Discovery deployed three satellites and removed ice particles from the orbiter using the Remote Manipulator System (the robotic arm). Resnik had developed operational procedures and software for the arm. In addition, she developed deployment procedures for a tether satellite system.

Resnik lived life to its fullest. She was a classical pianist and a gourmet cook. She was working on her pilot's license and liked to run and ride her bicycle.

SWE established the Resnik Challenger Medal in her memory. It is awarded for visionary contributions to space exploration. SWE also awards Resnik scholarships. The IEEE Judith A. Resnik Award recognizes outstanding contributions to space engineering [44, 45].

Bonnie Dunbar (1949–)

Astronaut Dr. Bonnie J. Dunbar is a pioneering engineering woman. When she enrolled as an engineering student at the University of Washington, there were nine women in her entire freshman class. She received B.S. and M.S. degrees in ceramic engineering from the University of Washington in 1971 and 1975, respectively. When she joined the Astronaut Corps in 1980, she was in only the second class at NASA to accept women. Subsequently, she earned her Ph.D. in mechanical/bio-medical engineering at the University of Houston in 1983.

Prior to becoming an astronaut, Dunbar was employed as a senior research engineer at Rockwell International Space Division, where she played a key role in the development of the ceramic tiles that form the heat shield for the space shuttle, allowing it to reenter the Earth's atmosphere. In 1978, Dunbar became a payload officer/flight controller for NASA. She served as a guidance and navigation officer/flight controller for the Skylab reentry mission in 1979.

While an astronaut, Dunbar's NASA technical assignments included: verification of shuttle flight software; serving as a member of the Flight Crew Equipment Control Board; 13 months in training in Star City, Russia for a 3-month flight on the Russian Space Station, Mir; and Assistant Director with a focus on University Research. She has logged more than 50 days in space.

Dunbar's experiments in space have involved protein crystal growth; surface tension physics; and tests on muscle performance, bones, the immune system and the cardio-pulmonary system. She received the Resnik Challenger Medal from SWE in 1992 and the IEEE Judith Resnik Award in 1993. She has been inducted into the Women in Technology International Hall of Fame [46, 47].

Eve Sprunt (1951–)

Eve Sprunt performed the fundamental work used (1) to identify oil and gas fields where there are naturally occurring fractures and (2) optimizing the resource recovery from those fields. Her work has led to the identification of areas where natural fractures are occurring, i.e., the identification of new oil and gas fields. With her work on seismic waves in rock, geologists can now predict the direction of the natural fractures in the rock in these fields. Once they know the direction of the natural fractures, they can fracture the rock to optimize resource recovery; the technique known today as hydraulic fracturing (also commonly called fracking).

Sprunt holds a B.S. and an M.S. in earth and planetary science, both from MIT. and a Ph.D. in geophysics from Stanford University as well as 23 patents. Sprunt's career was spent at Mobil Oil Corporation and Chevron Corporation. She served as the 2006 President of the Society of Petroleum Engineers and was the first woman to serve on the SPE board. During her term, she championed working with the United Nations to write common global standards for reserves and resource classifications. Sprunt founded and served as second president of the Society of Core Analysts (now the Society of Petrophysicists and Well Log Analysts). Sprunt led industry-wide collaborations that demonstrated the need for better quality control and the 10-year-long effort that revised the American Petroleum Institute's Recommended Practices for Core Analysis.

Sprunt received the 2013 SWE Achievement Award "For game-changing contributions to the petroleum industry, to the science and practice of geoscience and petroleum engineering, and to the advancement of women engineers" [48, 49].

Mae Jemison (1956–)

The first African-American female astronaut, Mae Jemison received her B.S. in chemical engineering from Stanford University. She then entered medical school at Cornell University Medical College. Jemison has global interests, studying in Kenya and Cuba during her medical school years and working at a refugee camp in

Thailand. Those global interests were enhanced during the 2 years she spent in the Peace Corps in Sierra Leone, after she obtained her M.D.

She pursued her dream of becoming an astronaut when she was selected as a member of the class of 1987. Jemison went into space in 1992 aboard the *Endeavor* where she conducted experiments on weightlessness and motion sickness on herself and her fellow crew members. Upon her return, Jemison remarked on how much women and other minorities can contribute if only given the opportunity. She has received many awards and recognitions.

Jemison taught at Dartmouth College and has established her own company, dedicated to encouraging a love of science in students and bringing advanced technology throughout the world. She has also established an international science camp.

Jemison said "I want to make sure we use all our talent, not just 25 %. Don't let anyone rob you of your imagination, your creativity, or your curiosity. It's your place in the world; it's your life. Go on and do all you can with it, and make it the life you want to live" [50, 51].

Kristina Johnson (1957–)

Inducted into the National Inventors Hall of Fame in 2015 and elected to the National Academy of Engineering in 2016, Dr. Kristina Johnson is the CEO of Cube Hydro, a company that advises in clean energy policy and invests in, develops and operates hydroelectric power facilities across North America. Prior to Enduring Hydro, Johnson served as Under Secretary of Energy at the U.S. Department of Energy. As Under Secretary, Dr. Johnson was responsible for unifying and managing a broad $10.5 billion Energy and Environment portfolio, including an additional $37 billion in energy and environment investments from the American Recovery and Reinvestment Act (ARRA).

Prior to joining the Department of Energy, Dr. Johnson served as Provost and Vice President for Academic Affairs at Johns Hopkins University, the largest research university in the U.S. From 1999 to 2007, Dr. Johnson was Dean of the Pratt School of Engineering at Duke University, the first woman to serve in that position. Before joining Duke University, Dr. Johnson served as a professor of electrical and computer engineering at the University of Colorado at Boulder, where she was a leader in interdisciplinary research on optoelectronics, a field that melds light with electronics. Her research and projects provided the University of Colorado approximately $42 million in grants and contracts. Her research and teaching included holography, which is the creation of three-dimensional images with light wave interference patterns, along with optical and signal processing, liquid crystal electro-optics and affixing a novel variety of liquid crystals to silicon to create new types of miniature displays and computer monitors.

In 1994, Johnson helped found the Colorado Advanced Technology Institute Center for Excellence in Optoelectronics. She also co-founded several companies including ColorLink Inc., KAJ, LLC, and Southeast Techinventures (STI). ColorLink makes color components for high definition television and other image projection devices utilizing the polarization, or vibrational, states of light. KAJ, LLC is an intellectual property licensing company that assists new firms using technology pioneered at the Optoelectronics Computing Systems Center at the University of Colorado at Boulder. STI is a technology acceleration company for commercializing intellectual property developed at Duke and other universities in the Southeast U.S.

In addition to her academic career, Johnson is an inventor and entrepreneur, holding over 45 U.S. patents (119 U.S. and international patents) and co-founder of several successful companies. Johnson has received numerous recognitions for her contributions to the field of engineering, entrepreneurship and innovation, including the John Fritz Medal, considered the highest award made in the engineering profession [49, 52].

Alma Martinez Fallon (1958–)

As Director of Supply Chain Procurement at Newport News Shipbuilding a Division of Huntington Ingalls Industries, Alma Martinez Fallon is responsible for approximately a $1 billion a year in material, subcontracting and service requirements for all programs at Newport News and joint procurement with General Dynamics Electric Boat.

Fallon began her career at Newport News in 1985 as a co-op student and then began her full-time career in 1988 as an engineer in the SEAWOLF Piping Engineering section of the SEAWOLF Engineering Division. She progressed to a Senior Engineer, Engineering Supervisor supporting numerous engineering design projects in the area of auxiliary piping and machinery systems for the Commercial Ship and Aircraft Carrier Programs. She was responsible for the planning for the *George H. W. Bush* aircraft carrier, the CVN-21 aircraft carrier, the Virginia Class construction programs and SAP/ERP3 for Steel Fabrication and Assembly. Prior to her appointment to Director, Fallon served as Hull Structure Construction Superintendent where she led the advanced planning, project management, design/build, and steel construction and assembly for the FORD Class.

Alma received a B. S. in Mechanical Engineering from Old Dominion University and a Master of Engineering Management from The George Washington University. She was born in the Dominican Republic and emigrated to the U.S. at the age of nine. She is bilingual with English as her second language.

A visible leader in the engineering profession, Fallon has a long and dedicated record of achievement in public policy and outreach. She was the 2004 national president of SWE and is a senior life member of the organization. She is an American Society of Mechanical Engineers (ASME) Fellow and she served on the ASME Board of Governors as a Governor (the first Hispanic so elected in ASME's history). Fallon also served as the 2007 Chair of the American Association of Engineering Societies.

She is the recipient of many awards including selection as one of America's leading minority women in technology by *Hispanic Engineer and Information Technology* magazine. Fallon was the Society Hispanic Professional Engineers 2004 Junipero Serra Award recipient. In 2012, she was the recipient of the *Inside Business* 2012 Women in Business Achievement Award [49, 53].

Ellen Ochoa (1958–)

The first female Hispanic astronaut, electrical engineer Ellen Ochoa became the Director of the Johnson Space Center in 2012. Selected for the astronaut program in 1990, she served on her first mission in 1993 aboard the space shuttle *Discovery*. She was in space on four separate occasions, logging over 1000 hours in flight.

Ochoa grew up in California, completing her undergraduate education at San Diego State University, earning a B.S. in physics and her graduate work in electrical engineering at Stanford University. She investigated optical systems for information processing and received three patents for an optical inspection system, an optical object recognition method, and a method for noise removal in images.

Ochoa is the recipient of numerous honors and awards including NASA's Outstanding Leadership Award and the Harvard Foundation Science Award [54, 55].

Sherita Ceasar (1959–)

Growing up in the "projects" (of the Chicago Housing Authority) may not have been the most auspicious start to an engineering career for Sherita Ceasar, but in high school, Ceasar heard about engineering while at a career fair. When the representative from the Illinois Institute of Technology (IIT) asked her if she wanted to make a lot of money after college, she was hooked. She attended an outreach program for minorities after her junior year in high school and placed second in mechanical aptitude out of 250 students. Ceasar was destined to study mechanical engineering.

After a B.S. and M.S. in mechanical engineering from IIT, Ceasar embarked on a career that has since led her to being the highest-ranking black female engineer within Motorola's Paging Products Group and is today a Vice President at Comcast.

As Director of Manufacturing at Motorola's Boynton Beach, Florida facility, Ceasar led an organization of nearly 2000 manufacturing associates, engineers, and managers in the manufacture of alphanumeric and numeric pagers. The facility was named by Arthur D. Little as the "Best of the Best in Manufacturing Management" and Ceasar represented Motorola at Arthur D. Little's "1995 Best of the Best Colloquium on Manufacturing Management."

A past National President of SWE, Ceasar received the 1997 Women of Color in Technology Award and has been inducted into the Women in Technology International Hall of Fame, in addition to numerous other awards. Ceasar personal motto is "I am a committed empowering leader who will make a difference in the world" [56, 57].

Padmasree Warrior (1961–)

Through 2015, the Strategic Advisor to Cisco (and formerly the Chief Technology Officer), Padmasree Warrior is trained as a chemical engineer, holding degrees from the Indian Institute of Technology and Cornell University. When she served as Senior Vice President, Engineering, Warrior was responsible for a wide variety of technologies including cloud computing core switching, and security. Before joining Cisco, Warrior was Executive Vice President and CTO at Motorola. During her time there, Motorola received the 2004 National Medal of Technology.

The recipient of numerous awards and recognition, Warrior was named by *Forbes* as one of "The World's 100 Most Power Women" for 2 years in a row. She has been inducted into the Women in Information Technology International Hall of Fame and named by *The Wall Street Journal* as one of the "50 Women to Watch."

A strong advocate for women in the tech industry, Warrior wrote in the *Huffington Post* "the fact that you're different and that you're noticed, because there are few of us in the tech industry, is something you can leverage as an advantage." She has advised women to take opportunities as they arise and not second guess their own capabilities. Warrior also advised women to focus on work-life integration as opposed to work-life balance [58, 59].

Kalpana Chawla (1962–2003)

The first Indian-American woman astronaut and the first Indian-American woman in space, Dr. Kalpana Chawla received her undergraduate aeronautical engineering education in India and her graduate aerospace engineering education in the U.S. Her early work was in the area of powered-lift computational fluid dynamics. She researched complex air flows around aircraft.

Selected for the astronaut program in 1994, her training began in 1995. She flew aboard the space shuttle *Columbia* in 1997 whose mission was a Microgravity Payload flight that focused on experiments designed to study how the weightless environment of space affects various physical processes, and on observations of the Sun's outer atmospheric layers. Her second flight aboard *Columbia*, which was a research and science mission involving 80 experiments, ended in tragedy in 2003 [60].

Sandra Begay-Campbell (1963–)

A Principal Member of the Technical Staff at Sandia National Laboratories, Sandra Begay-Campbell leads the laboratories' efforts to assist Native American tribes with their renewable energy endeavors. A civil engineering graduate of the University of New Mexico and Stanford University, Begay-Campbell grew up in the Navajo Nation fascinated by math and science and trying to figure out how things work.

The recipient of numerous awards and honors, Begay-Campbell says she enjoys working in the renewable energy field because "it's wide open and cutting edge." She has been very active in the American Indian Science and Engineering Society and was the first woman to chair its Board of Directors [61–63].

Kristi Anseth (1968–)

A pioneer in the field of biomedical engineering, a Distinguished Professor at the University of Colorado and a member of both the National Academy of Engineering and the National Academy of Sciences, Kristi Anseth is a leading researcher and inventor in the fields of biomaterials and regenerative medicine. Her work has demonstrated that by controlling the chemical, biological, and physical properties of biomaterials, fundamental cell biology issues can be determined and the information so obtained can be used to regenerate tissue. The results of her work mean that broken bones heal faster and diseased heart valves can be replaced. Anseth's seminal work has revolutionized the field resulting in biomaterials that are tissue substitutes that are able to restore, maintain, or improve tissue function.

The first engineer to be named a Howard Hughes Medical Institute investigator, Anseth has been named one of the "100 Chemical Engineers of the Modern Era." A dedicated teacher who mentors and promotes the careers of her students, Anseth serves on federal review panels including those at the National Institutes of Health. Her degrees are in chemical engineering from Purdue University and the University of Colorado [64–66].

References

1. Lucena JC. "Women in engineering" a history and politics of a struggle in the making of a statistical category. Proceedings of the 1999 international symposium on technology and society—women and technology: historical, societal, and professional perspectives, pp 185–194. New Brunswick, NJ, Accessed 29–31 July 1999
2. National Science Foundation (1996) Women, minorities, and persons with disabilities in science and engineering: 1996, Arlington, VA, (NSF 96-311). The quote is from the Science and Engineering Equal Opportunity Act, Section 32(b), Part B of P.L. 96-516, 94 Stat. 3010, as amended by P.L. 99-159
3. Engineering Workforce Commission of the American Association of Engineering Societies (1998) For Engineering Education, 1997 Outputs Look Like 1996. Engineers. 4(1):12
4. (1992) The glass ceiling & women in engineering. Report of the NSPE Women in Engineering Task Force. NSPE Publication, Alexandria, VA
5. Tobias S (1997) Faces of feminism: an activist's reflections on the women's movement. Westview Press, Boulder, CO, p 115
6. Engineering Workforce Commission of the American Association of Engineering Societies (1999) Engineering degree totals slump. Engineers. 5(4)
7. National Academy of Engineering (1999) The summit on women in engineering. Program Book, Washington, DC

8. Engineer Girl. www.engineergirl.org. Accessed 6 June 2015
9. (2000) Land of plenty. Report of the Congressional Commission on the Advancement of Women and Minorities in Science, Engineering and Technology Development
10. NACME: our mission & strategy. www.nacme.org/org.html. Accessed 19 Feb 2001
11. NAMEPA, Inc.: mission statement. www.namepa.org. Accessed 19 Feb 2001
12. Bill and Melinda Gates Announce New Millenium scholars program to bridge the gap in access to higher education. 16 Sept 1999. www.techresource.org/press/990916statement.html. Accessed 28 Oct 1999
13. National Academy of Engineering, Greatest Engineering Achievements of the 20th Century. http://www.greatachievements.org/. Accessed 2 June 2015
14. National Academy of Engineering (2004) The engineer of 2020: visions of engineering in the new century. The National Academies Press, Washington, DC
15. (2010) Table 22. Life expectancy at birth, at 65 years of age, and at 75 years of age, by race and sex: United States, selected years 1900–2007. http://www.cdc.gov/nchs/data/hus/2010/022.pdf. Health, United States. Accessed 13 June 2015
16. Horowitz S. Wal-Mart to the rescue: private enterprise's response to Hurricane Katrina. Independent Rev. https://www.independent.org/pdf/tir/tir_13_04_3_horwitz.pdf. Accessed 13 June 2015
17. American Society of Civil Engineers (2013) Report Card for America's Infrastructure. www.infrastructurereportcard.org. Accessed 2 June 2015
18. Tietjen JS (2008) Honoring the Legacy of Ada Pressman, P.E. SWE Mag Soc Women Eng. Fall
19. Pressman AI. Encyclopedia of world scientists, revised edition, Accessed 6 June 2015. http://www.fofweb.com/History/HistRefMain.asp?iPin=EWSR0723&SID=2&DatabaseName=American+History+Online&InputText=%22California%22&SearchStyle=&dTitle=Pressman%2C+Ada+Irene&TabRecordType=All+Records&BioCountPass=2925&SubCountPass=2285&DocCountPass=365&ImgCountPass=
20. Sheila Widnall. p. 1–2. www.witi.com/center/witimuseum/halloffame/1996/dwidnall.shtml. Accessed 22 Dec 2000
21. Widnall SE. www.wic.org/bio/swidnall.htm. Accessed 22 Dec 2000
22. Widnall SE. pp. 1–2. www.af.mil/news/biographies/widnall_se.html. Accessed 22 Dec 2000
23. (1998) 1998 IEEE medals. IEEE Spectrum. 66
24. Ambrose SA, Dunkle KL, Lazarus BB, Nair I, Harkus DA (1997) Journeys of women in science and engineering: no universal constants. Temple University Press, Philadelphia, pp 422–425
25. Flowers SH, Abbott MH. Women in aviation and space, U.S. Department of Transportation Federal Aviation Administration, undated, p 8
26. Stanley A (1995) Mothers and daughters of invention: notes for a revised history of technology. Rutgers University Press, New Brunswick, NJ, p 408
27. (1995) Who's who in technology, 7th edn., Gale Research, Inc., New York, p 1328
28. Veronique L. (2015) Interview: Mary-Dell Chilton on her pioneering work on GMO crops, Genetic Literary Project, St. Louis Public Radio. http://www.geneticliteracyproject.org/2015/05/27/interview-mary-dell-chilton-on-her-pioneering-work-on-gmo-crops/. Accessed 6 June 2015
29. National Inventors Hall of Fame, Inductees: Mary-Dell Chilton, http://invent.org/inductees/chilton-mary-dell/. Accessed 6 June 2015
30. (2015) The World Food Prize, Syngenta Scientist Dr. Mary-Dell Chilton Named 2015 National Inventors Hall of Fame Inductee. http://www.worldfoodprize.org/index.cfm/24667/35489/syngenta_scientist_dr_marydell_chilton_named_2015_national_inventors_hall_of_fame_inductee. Accessed 6 June 2015
31. Dr. Eleanor Baum. www.witi.com/center/witimuseum/halloffame/previousinducte/1996/dbaum.shtml. Accessed 1 July 1999, pp 1–2
32. Why outward bound for engineers? p. 1. www.cooper.edu/engineering/projects/outward-bound/obstory1.html. Accessed 30 Dec 1998

33. (1993) Eleanor Baum. IEEE Spectrum. pp 42–44
34. (1991) Rebel with a cause becomes 1st lady dean of engineering. Natl Enquirer 10
35. Managing creativity: Donna Lee Shirley http://www.managingcreativity.com/shirley.html. Accessed 6 June 2015
36. Shirley D, Morton D (1998) Managing Martians. Broadway Books, New York. This book is subtitled, "The extraordinary story of a woman's lifelong quest to 'get to Mars'—and of the team behind the space robot that has captured the imagination of the world"
37. Profiles of Women at JPL: Donna Shirley. www.jpl.nasa.gov/tours/women/Shirley.html. Accessed 22 Dec 2000
38. de Planque GE (2014) Nomination to the Maryland women's hall of fame, in the files of the author. Unpublished
39. Proffitt P (ed) (1999) Notable women scientists. Gale Group, Farmington Hills, MI
40. First Lady Hillary Rodham Clinton to Speak at Inaugural Gala for Rensselaer's 18th President, The Honorable Dr. Shirley Ann Jackson, Press Release, September 17, 1999, www.rpi.edu/dept/NewsComm/New_president/presshillary.htm. Accessed 23 Nov 1999
41. Perusek AM (2002) Saluting African Americans in The National Academy of Engineering Class of 2001. SWE: Mag Soc Women Eng 24–26
42. "Beyond 2000: exploring perspectives," Society of women engineers national conference, program book, June 27–July 1, 2000, p. 47. 2000 SWE achievement award nomination package for Suzanne Jenniches, Accessed 15 Dec 1999
43. Suzanne Jenniches F. Nomination for the Kate Gleason Award, in the files of the author. Unpublished
44. Judith A. Resnik Award, swww2.ieee.org/about/awards/sums/resnik.htm. Accessed 14 Feb 2001
45. Brody S. "Judith Resnik," pp 1–2. www.us-israel.org/jsource/biography/Resnik.html. Accessed 14 Feb 2001
46. Dunbar BJ. pp 1–2. www.witi.com/center/witimuseum/halloffame/2000/bdunbar.shtml. Accessed 22 December 2000
47. Biographical data. pp 1–2. www.jsc.nasa.gov/Bios/htmlbios/dunbar.html. Accessed 22 Dec 2000
48. Achievement Award, Eve Sprunt, Ph.D., Chevron Corporation, SWE Magazine, Conference (2013) http://societyofwomenengineers.swe.org/images/AwardRecipients/Achievement/Achievement_Award_EveSprunt.pdf. Accessed 6 June 2015
49. Unpublished nomination in the files of the author
50. Jemison MC. Biography: astronaut, doctor. http://www.biography.com/people/mae-c-jemison-9542378. Accessed 6 June 2015
51. Redd NT (2012) Mae Jemison: astronaut biography. http://www.space. Accessed 6 June 2015
52. Johnson KM. National Inventors Hall of Fame. http://invent.org/inductees/johnson-kristina/. Accessed 6 June 2015
53. BridgeBizSTEM, Brown, Vi, Her-Story: Alma Martinez Fallon, https://bridgebizstem.wordpress.com/2014/07/26/her-story-alma-martinez-fallon-2/. Accessed 6 June 2015
54. Ellen Ochoa. Biography: engineer, astronaut. http://www.biography.com/people/ellen-ochoa-10413023. Accessed 6 June 2015
55. NASA. Johnson Space Center Director Dr. Ellen Ochoa, http://www.nasa.gov/centers/johnson/about/people/orgs/bios/ochoa.html. Accessed 6 June 2015
56. Tietjen JS, Schloss KA, Carter C, Bishop J, Lyman Kravits S (2001) Keys to engineering success. Prentice Hall, Upper Saddle River, NJ, p 25
57. Ceasar ST. pp 1–2. www.witi.com/center/witimuseum/halloffame/1999/sceasar.shtml. Accessed 22 Dec 2000
58. The Network, Cisco Technology News Site, Padmasree Warrior, Strategic Advisor, http://newsroom.cisco.com/padmasree-warrior;jsessionid=BC304CAA97923359D4F0316F06729ACE?articleId=34150. Accessed 6 June 2015
59. Quinn M (2015) Quinn: women in tech lose a high-placed warrior. San Jose Mercury News. http://www.mercurynews.com/michelle-quinn/ci_28254424/quinn-women-tech-lose-high--placed-warrior. Accessed 6 June 2015

60. NASA Biographical Data Sheet for Kalpana Chawla. http://www.jsc.nasa.gov/Bios/htmlbios/chawla.html. Accessed 22 Aug 2015

61. Sheppard LM. Portfolio: profile & biographies: an interview with Mary Ross. Lash Publications International. http://www.nn.net/lash/maryross.htm. Accessed 6 June 2015

62. Short biography: Sandra Begay-Campbell. http://globals.federallabs.org/pdf/bio_begay-campbell.pdf. Accessed 6 June 2015

63. Real scientists: Sandra Begay-Campbell. http://pbskids.org/dragonflytv/scientists/scientist55.html. Accessed 6 June 2015

64. Anseth KS. Chem Biol Eng. University of Colorado Boulder. http://www.colorado.edu/chbe/kristi-s-anseth. Accessed 6 June 2015

65. Anseth K. Colorado women's hall of fame. Biomed Eng http://www.cogreatwomen.org/index.php/item/172-kristi-anseth-phd. Accessed 6 June 2015

66. Anseth KS. Purdue College of Engineering, Our People, https://engineering.purdue.edu/Engr/People/Awards/Institutional/DEA/DEA_2012/Anseth. Accessed 6 June 2015

Index